Francois Louis Schauermann

Theory and Analysis of Ornament

Applied to the Work of Elementary and Technical Schools

Francois Louis Schauermann

Theory and Analysis of Ornament
Applied to the Work of Elementary and Technical Schools

ISBN/EAN: 9783744644709

Printed in Europe, USA, Canada, Australia, Japan

Cover: Foto ©berggeist007 / pixelio.de

More available books at **www.hansebooks.com**

THEORY AND ANALYSIS

OF

ORNAMENT

APPLIED TO THE WORK OF

ELEMENTARY AND TECHNICAL SCHOOLS

BY

FRANÇOIS LOUIS SCHAUERMANN

For Eight Years Head Master of the Wood and Carving Department,
Royal Polytechnic, Regent Street

WITH 733 DIAGRAMS AND ILLUSTRATIONS

LONDON
SAMPSON LOW, MARSTON & COMPANY
Limited
St. Dunstan's House
FETTER LANE, FLEET STREET, E.C.
1892

LONDON :
PRINTED BY GILBERT AND RIVINGTON, LIMITED.
ST. JOHN'S HOUSE, CLERKENWELL ROAD, E.C.

PREFACE.

This book has been compiled because of a strongly felt want of a complete text-book upon the subject of Theory and Analysis of Ornament, the Author having been asked by his own and other students, why he did not write a book more in advance of the teaching of the present time. The study of Ornament has made such rapid progress during the last twelve years, that those books which were previously quite efficient have become obsolete. So far as the Author's knowledge extended, there seemed a very large demand for books of an explanatory nature, as applied to the work of elementary and technical schools. This work is published with the object of more completely developing the various sections of the subject hitherto dealt with. The arrangement is such as to present the entire subject in the simplest and most easily to be understood form.

The present Minute of the Education Department on Science and Art, encouraging technical work in the schools, promises great results ; and the Author has endeavoured to produce a book of practical use to teachers, in their preparation for the tuition and examination of schools.

FRANÇOIS LOUIS SCHAUERMANN.

CONTENTS.

LIST OF ILLUSTRATIONS.

THEORY AND ANALYSIS OF ORNAMENT.

Part I.

CHAPTER I.

1. Diapered forms are productions of the artist, and are in contradistinction to those produced by nature. Forms produced by nature bear the mark of a first creative power; whereas, the diapered forms, produced upon those original forms, are the productions of the human brain assisted by certain tools employed, which are themselves in turn the productions of mechanical skill. The two things necessary to bring about or carry out the artist's work, are material and industry. Industry is, then, the " ways and means" by which the artist brings about the productions of his art. Nature furnishes the raw material, art produces form, in the raw material, necessary to the wants of man, both artistic and useful.

2. The artist's work can easily be distinguished from that which shows mere mechanical skill. Mechanical skill can be recognized as proceeding from practice in a certain art or industry; the artist's work indicates a higher inherent power, but can easily be distinguished from an untutored production of the brain.

B

The diapered form, independently of its beauty and usefulness, is a compound of three elements, viz., material to be used, mode of manipulation, and, thirdly, its relation to art.

The material to be worked furnishes its plastic and decorative qualities.

The types depending on art are form and decoration. Form is the result of particular modifications wrought in the material by the workman, and is necessary to the satisfying of the æsthetic taste of the worker. Decoration is an addition made to the first form, as produced by the artisan, and is the work of the artist.

The following table should be carefully studied, as giving the principal forms and types as brought into existence by the artist or mechanic. The mechanic's work must be acknowledged as a very necessary adjunct to the complete production of the diapered forms produced by the inborn genius of the artist himself.

FORMS.

(i.)	Modelled Forms. (Plastic.)	Modelled (plastic). Turned (ceramic). Blown (glass works). Hammered (brass ware).
(ii.)	Carved Forms. (Sculpture.)	Carving (carving). Sculptured (sculpture). Chiselled (toreutic). Engraved (glyptic).

(iii.) Assembled forms (joinery, iron-foundry, &c.).
(iv.) Built or added forms (construction, architecture).
(v.) Weaved forms (weaver's trade, basket trade).

DIAPERED FORMS.

3. Materials produced by nature are peculiarly

adapted to the manipulation of the artist, and the latent qualities afford the genius in man an opportunity to exercise its powers of producing decorative forms and types so pleasing to the mind and necessary to the true enjoyment of nature.

Man sees the raw productions of nature, and, concurrently with sight, springs up a desire to beautify, which desire calls forth an inherent talent capable of being more highly developed in those whose natural qualifications are those of an artist apart from or above those of the mechanic.

The following table should be carefully studied, as giving the different modes of decoration.

(I.) The illuminated, painted or enamelled decoration; which refers to the modelled, carved, assembled, builded and weaved, diapered forms.

(II.) The modelled or sculptured decoration; which refers to the first four, viz., modelled, carved, assembled and constructed forms.

(III.) The engraved and carved decoration; which refers to modelled, carved and assembled forms.

(IV.) The embroidered decoration, which refers exclusively to the weaved forms.

ORNAMENTATION.

4. Form and decoration are so closely allied that the one naturally forms part of the other; that the blending of the two makes one quality, viz., that of ornamentation. Under the generic name of ornament, we must consider the numberless variety of drawings, forms, reliefs, coloured prints, &c., by which man has surrounded himself, and which bear testimony to his ingenuity and inventive power.

It must be borne in mind that ornament is purely the production of art, and an evident distinction exists between the result of one man's genius and that which comes of the acquired skill of another. The artisan can produce a diapered form but the artist's genius is first called for as a necessity for producing the model for that form. Both the artist's and artisan's work are products of industry; the one has, for its success, a stimulant in the inherent and instinctive desire to produce; the other plies his industry as a means of living by an art acquired by practice, and the few improvements that must follow a continued exercise of his craft.

5. The Divine power of the Creator is apparent in that He supplies the rough material in such forms that it is easily worked upon for the production of artistic forms, as a result of a special artistic power given by the same great Donor. Form and ornamentation have ever been a means for pleasure; and the products of art have formed the basis for much of the success or failure of men who have constituted themselves teachers or leaders, from time to time, in the world's history. Great leaders in religious matters were, in the Middle Ages, among the best producers in carving and works of a kindred art, and much of their success is traced from their ability to imitate nature and to diaper imaginary forms for the embellishment of their monasteries or religious houses and other buildings. The artistic powers of the Middle Ages are abundantly manifested in cathedrals and churches built at that age, the decorations being principally the productions of those monks possessed of artistic ability, and who were almost the sole representatives of the refining influences of education. In the Middle Ages the artisan was employed for carrying out the work of genius as dictated by it, and it is no less now the labour of the mechanic that supplies the apparatus for the artist's use, by which he can transform nature and convey correctly the productive forms of one country to another, however far apart upon the face of the earth.

Nature has two distinctive characteristics; those perceptive and those conceptive—those in actual being and those that we can conceive as possible, to understand which, we must conceive four worlds.

1. Existing world (the actual universe).
2. Historical world (world of names and events).
3. Fabulous world (mythological world).
4. Ideal world (imaginary world).

To the *existing world* belongs the actual universe, the physical formations, moral and political governments, &c.

The *historical world* is that which records events and tells of the great movers in those events, with their epochs.

The *fabulous world* is that which is full of imaginary gods, heroes and heroines.

The *ideal world* is that where beings exist only in imagination, and who are, with imaginary objects, treated as having real existence.

The artisan or the artist has a limited horizon, he must confine himself and his art to things

strictly terrestrial, leaving for the philosopher a
separate field for the exercise of his learning
and study.   The artist must have reality in his
model, that is to say, his conceptive powers must
be limited to those objects that can be imagined as
forming a part or parts of the existing universe,
yet his imaginative powers must not be so much
circumscribed as to injure his form, type or picture.
He must leave telescopic worlds and microscopic
nature free from his portrait or diapered form or
type.

The theologian, the moralist, the poet and other
learned men may treat of the actual universe, and
yet seek further by the aid of telescope, microscope
and imagination to embellish their works; but in
material forms the true artist will confine himself
to the copy of real nature, and any introduction to
his world must be actual and not imaginative phenomena.   The painter artist has a world even less
limited than that of the sculptor or carver.   The
painter finds many of his finest touches from the
clouds as governed and tinted by the varied conditions of the atmosphere; but as the sculptor or
carver cannot diaper light or colour, his world is
necessarily more limited than that of the landscape
painter.

6. Form may really be said to have a threefold
existence, viz., natural, diapered and æsthetic : the
natural form being that in which nature produced
it ; the diapered form a result of the artist's skill ;
and the æsthetic form that produced under varied
degrees according to the imaginative powers of the
artist.

The technical mode of decoration must be strictly applied in ornament for producing a thorough harmony ; ornament wrongly or carelessly applied to the diapered forms produces discord, and is therefore repulsive to the æsthetic eye of the art critic. The æsthetic eye sees in the natural form the diapered form or type produced, and fancy adds ornamentation to be carried into effect by the industrial application of skill. The natural artist, upon first contemplating original form, perceives every geometrical figure, geometry forming, as it were, a most important constituent of his brain-power ; therefore geometry must be cultivated for the complete reproduction of objects of nature in diapered forms or types.

7. As all objects in nature possess peculiar powers of motion under the heads generation, progression, concatenation, variation, &c., the artist's mind must associate these powers with geometry as necessary to the successful delineation of his work. Suppose for instance, that the artist diaper the form of a horse ; for the true reproduction, his diapered form must bear out the notion of movement, and the prominent features of the diapered form must convey an idea of the exercise of the five senses.

The abstract, rational science called Kinematics, is the power of transforming or reproducing, in the diapered form, the true compositions of varied motion. Thus the idea of motion, in the abstract, enters into combination with geometry in the mind of the artist.

The idea of motion, independent of the

notion of actual movement, combines with geometry, and this combination being correctly delineated produces, in the diapered form, the nearest approach to actual life. In short, all the attributes of life must be as correctly copied as possible, and be projected as prominently in the figure as art can produce in the way of actual creation.

The ideas of motion and swiftness must not be associated with the idea of strength. The practical kinematic plies his skill upon the basis of motion and swiftness, but it remains for the genius of the artist to convey the notion of existing strength. Thus, the idea of equilibrium, separate from the ideas of strength and motion, combines with æstheticism, and shows by the necessary use of the words balancing, ponderation, proportion, &c., the æsthetic determination of forms.

The idea of equilibrium, apart from the idea of motion but associated with the idea of strength, or rather the idea of occult and permanent endeavour or stability, is the basis of operation by the mechanical architect. The idea of disposable or exhaustive strength is the basis of operation by the mechanical engineer.

It must be realized by the student, that the idea of strength hides itself behind higher ideas which become indefinite in ordinary logic as produced by scientific construction. Where language is employed to explain, scientific construction conveys, by expressive metaphors, the idea of strength, activity, attraction, intensity, energy, &c.

The following table must be studied in order fully to grasp the meaning of the foregoing remarks.

PURE GEOMETRY.

Analytic geometry (Kinematics) = The artisan's mechanics.

Descriptive geometry (Statics) = The architect's mechanics.

Dimensive geometry (Dynamics) = The engineer's mechanics.

8. Every motion, instinctive or derived, contains two or more elementary parts, viz., simple motion and movement in straight or circular lines, which may really be reduced to two simple motions, or technically speaking, Helicoidal and Volubilial motions.

The three geometric elements, having the property of constant juxtaposition (slippery relief on incrustation), are, the straight line, the circular line and the spiral line.

The straight line, the circular and spiral lines, are the fundamental, simple and irreducible elements of geometry.

The rectilinear, circular and helicoidal motions are the three fundamental, simple and irreducible elements of kinematic or theoretical geometry as applied to motion.

The two rectilinear motions, circular and translatory, can be continuous or alternative, i.e., constantly directed one way, or alternately directed over a plane surface; thus, from the four elementary motions, which form the basis of industry in the kinematic or artisan's mechanics, the simplest as well as the most complicated engine can be made to transmit or transform.

9. We have spoken of simple and uniform motions to which are corresponding simple and uniform geometric lines, viz., straight, circular and spiral lines.

Motion can be varied, i.e., accelerated or slackened, or caused to follow, in continuation, the three fundamental lines, though not to deviate from the geometric figuration of motion. The notion of swiftness can be formed as the result of time employed and space traversed. Motion, then, can be said to be mathematical and special, and although mathematically simple in itself, is yet quite strange and foreign to the æsthetic ideas of order and form.

Form, the geometrical or abstract, is, therefore, shown to be the result of motion.

We will now see, or deduce from the foregoing remarks, why mathematicians find a place in the world of art.

10. A study of geometry is very necessary to art, as it gives a comprehensive and correct idea of form, motion and space, both in the actual and the abstract. By geometry the artist is able to apply the abstract forms, space and motion, necessary to the complete production of his work, either of the brush or of the chisel.

There are two kinds of geometry, viz., Dimensive and Descriptive. By the aid of dimensive geometry, lengths, breadths and volumes may be estimated. By the aid of descriptive geometry, lines, surfaces, specifications, &c., together with motion upon planes, may be estimated with great exactness.

Among the different properties of figurative extent are lines and surfaces geometrically determined alike in plane surfaces as in solids, the geometrical lines being employed in every case as determining the extent or volume of the figure in question.

These two kinds of geometry may again be classed as elementary or euclidian, and modern or general geometry. These divisions in geometry must be recognized in order that the student may grasp the full reason why science must be applied to art, mechanical or inherent, for the full development in production.

11. Now, instead of accepting, indefinitely, the dogmatic as adhered to, generally, by teachers and lecturers, we will consider the subject as a whole. A full acquaintance with the following paragraphs will assist the student in grasping the meaning without encumbering his mind with the sub-division.

Four heads : Analytic, Synthetic, Descriptive, Dimensive.

Pure geometry, as established by the Greeks and adopted as classical in our schools, places before the eye of the student, logical strictness and perfection : it teaches demonstration, concatenation, and subordination of proportion; it teaches also the laws of definition and classification geometrically considered. Such logical foundations constitute the real establishment of geometry as a science.

Geometry, studied under the four heads given

above, leads the mind to admiration of the broad
field for expressed exactness opened by its study.
No science is more adapted to the establishment of
logic, the science being logic itself, and where logic
exists, the mind endowed with it must manifest its
capability for applying it to objects more or less
artistic. The practical artist is, in his works, an
illustration of the useful application of pure
geometry.

ANALYTICAL GEOMETRY.

The mathematician applies the science of algebra
to the working out and proving of the truth of
problems in geometry, and consequently truths are,
by analytical geometry, proved, as relating to
questions based upon the science of geometry, and
to which that science must be applied for perfect
elucidation.

Descriptive geometry is the application of special
and artificial methods, viz., the methods of projections, as applied to frames and solids.

By descriptive geometry we understand the theory
of intrinsic composition of forms, and the definition of different particularities of the lines and
surfaces. This statistical contemplation of the
forms of extent is not at all subordinated to a
particular drawing method, nor to a spiral ter-
minology.

The geometric, like all other forms, might be
described by ordinary language, and might be drawn
up by the usual artifices of the drawing art.

The ideas of generation and description are so
closely allied that analysis does not separate the

ideas advanced under each head, yet it is necessary
to distinguish the two orders of perceptions, as the
one goes to the root of the science and separates
the parts so as to give explanation to their generic
independence, the other explains the simultaneous
presence of their dispositions. Perhaps the distinction is vague, and the language employed must,
of necessity, be somewhat vague also; yet a study
of the matter will throw light upon the somewhat
clouded expressions, for geometry is a science well
adapted to prove its genuineness in its own un-
uttered language. When the science of geometry
is applied, as it must be, to the true production of
nature, the artist may apply under the several sub-
divisions, and yet keep aloof from the idioms or
rubrics of the science, simply applying the abstract
to the correctness of his art.

12. Geometry, as taught by mathematicians, forms
an important branch of the general system of mathematics, which contains the three following theories:—

1. The theory of numbers.
2. The theory of figures.
3. The theory of sizes.

Pure mathematics are absolute and rational, and
most perfect examples of form and construction, and
form the secret of pre-eminence in the scientific
art-roll.

The perfection of the mathematician has proved
his license to intrusion into the artistic world, either
to the damage or benefit of the latter (?), and the
exactness of his calculations presents a formidable
critique upon the work of the artist and compels him
to stoop to criticism.

13. The teacher of geometry employs, of necessity, a rude method of drawing, and the science he teaches admits of dogmatic sanction of the clumsy and inartistic method called linear drawing. Linear drawing may be classed as an intrusion upon the artist's freehold. The student in geometry requires ruler and compass, and upon their use depends the success of his linear drawing; the artist supplies these necessities by the natural and artistic effort of his brain.

The student in mathematical geometry can, with the help of the ruler and compass, produce neatly-effected work, yet he must admit that his productions are chiefly the result of the instruments he employs.

The careful student in geometry carries, in his work, the power of imposing upon the unartistic world—he is deemed, by a credulous public, an artist, whereas his productions are by no means resulting from artistic genius. Until art has passed through, or proved itself above this sham, the artist must be in constant rivalry with the pretentious linear drawing student.

That the cause of true art must ultimately triumph is certain : as a consequence, theoretical education in art must assert superiority over those who have passed by the preliminary education and taken their first step half-way up the ladder, from which they cannot rise nor will their bigotry admit of descent.

The student will perceive that the science of geometry is only shown in its true beauty when applied to art productions.

14. Science may be said to be knowledge logically organized, and logical organization is based on the concatenation of propositions. Logic reduces the initial notions as much as possible, thus covering much important matter and difficult questions under a few words, questions the solutions of which cannot be arrived at but by logical argument. Those geometrical propositions, logically formed, can be stated thus :—

1. To reduce the fundamental notions as much as possible under the dogmatic form of axioms, definitions and postulates.

2. To satisfy logical rigour by giving demonstration to all that is susceptible of it.

All geometrical teaching is founded upon the foregoing method, assisted by numerical calculation, applied to geometrical sizes, and propositions in illustration.

Geometry, as a science, does not admit of exercise in what has been described as " Mathematical Joy." The logic applied, and the application required in the solution of its problems, take up the whole attention of the student, who must bend his efforts to his work, leaving no spare time or energy for devotion to the work of pure art ; in other words, did artistic ability exist, it is destroyed or negatived by the arduous work of the student in geometry.

Before the intervention of measuring and counting and mechanical linear drawing, the first principles of geometry were acknowledged to be abstract ideas, which predominate in all branches of mathematics. True artistic study can produce the same result, although the artist does not, for he must not undertake to plunge too deeply into the maze of abstract ideas, figures or measurements; his mind supplies spontaneously much that the geometrical student acquires by hard study. In short, the student in geometry is one being, and the artist another.

15. The materials, of the science of geometry, are sizes or mathematical quantities, conceived in all simplicity, classified and constructed upon the general principles of order and form.

If, in the attempt to study art, the student in geometry isolates the susceptible forms or figures of mathematical propositions, he leaves the world of human conditions, and plunges into a maze of logical abstractions. The attempt results in failure.

The work of the artisan preceded the science of geometry, therefore, Science may be called "a knowledge of knowing." Its place is in the rear of poetry, art, history, moral and philosophical compositions, thus showing that science is only regulated argument upon facts established. Science cannot be maintained but by profound study of truths recognized, and following the rules strictly laid down under the terms assiduous and disciplined application.

16. Geometry may be looked upon as suddenly springing up from a latent existence, and as suddenly presuming to govern the arts. The presumption is, that as art existed without practical geometry, science presumes a right to govern the arts.

All that was produced by moral activity, now forms objects for special and necessary culture. Methods and systems have been arranged to properly develop the intellectual faculties, and the earlier the mind is called upon to learn those methods and systems so as to become acquainted with the field, the sooner can the matured mind be brought to grapple with the facts hidden underneath. Science, then, developed by education and intellect; it is inborn in man, and its full development depends upon the strength of brain-power assisted by application; and the more advanced one becomes in scientific knowledge, the further away does he remove himself from the real sentiments of life. The scientific mind sees order and neatness in everything in creation, it detects disorder in everything that has passed under manipulation; and although removed from the real sentiments of life, it finds a world of thought and beautiful reverie as a result of study and appreciation of the beautiful which is always present in God's creation. This, then, is the rule of science, and under this rule there is no place but for organized activity. The humble artisan and the busy mechanic must submit themselves to the guidance of men who are scientifically capable of directing their efforts.

However distasteful to the worker, such a situation must be considered as an unavoidable condition of modern life. The art student goes into the past for his lessons, he takes the great masters of by-gone ages for his guides, his deductions from their works are brought into actual contact with

and made to bear upon modern ideas; and
such a blending of the past and present cannot
but be productive of greater perfection, and
that this perfection should reign supreme should
not be looked upon with dissatisfaction, but rather
rejoicing, seeing that science adds to the com-
fort and blessings of all living under its bene-
ficent sway. When we extol the cause of this, we
must not be led away to forgetfulness that much
is due to the artisan, much improvement has been
made as the result of skill, exercise and persever-
ance, and, in many cases, great inventions have been
made by the mechanic; yet it must be acknowledged
that where the mechanic's mind has attained to
such scientific perfection, he has been assisted by
an inborn artistic nature which would have quali-
fied him for a place among the artists, had he
been allowed to follow the dictates of nature
rather than been led off to a mere mechanical
exercise of his faculties. The world cannot rest,
science must have its sway, and however much the
mechanical mind revolts, its masters and best pro-
fessors of mechanical skill must acknowledge the
power of the scientist. Should it be so? Yes,
unless pride bars the advance of progress so necessary for meeting the enormous requirements of a
thickly-populated world.

CHAPTER II.

ORDER, FORM, RELIEF, COLOUR.

17. The idea of form corresponds with the
natural, diapered or inventional, which have three

dimensions. We will call the " Fixing of Limits,"
" Extension of Form; " the figure ending in outlines or lines of delineation. This figure is the
representation; the image or the drawing of a
diapered form, or it may be termed a purely inventional figuration.

18. The idea of order governs the constitutive
parts of form, that is, it implies a diagram or
plan of disposition of purely intelligible type of real
form.

The idea of order is more particularly the idea of
arrangement of parts distinguished only by analysis ;
it constitutes the disposition of æsthetic form and
ornament, and consequently the law of monumental art.

19. The notion of Form has two acceptations.

1. The abstract, figurative and geometrical
form.

2. The æsthetic or concrete and the expressive
form.

All particular forms are diapered or described—

1. By geometrical types based upon terms of
geometrical exactitude.

2. By picturesque types, for the definitions of
which are employed numerous arbitrary metaphors,
which explain forms in nature as well as in art.

The idea of abstract form, at first purely geometrical, contains particular points of view :

1. Material, or corporeal forms.

2. Ideal, configurative and enveloped forms,
which embrace the most showy points of material,
organized or diapered form.

The notion of concrete form contains the material,

it also contains the special points of view of organized form, of diapered and, in fact, all æsthetical form.

20. The following table arranges the different accepted notions of form and shows their order of mutual dependence :—

ÆSTHETIC FORM.

Monumental, Plastical, Picturesque, Decorative, Ornamental.

Diapered Forms.	*Organized Form.*
Assembled.	(Morphology of
Built.	Animals and
Weaved.	Vegetables.)
Material.	Envelope Form.
Carved.	Diapered Form.
Modelled.	(Organized Form.)
(Minerals.)	Material Forms.

ABSTRACT FORMS.

21. The abstract form is more particularly geometrical, when it can be defined by the rigorous methods of mathematical science. The abstract plan of construction must be distinguished from the concrete or æsthetical.

The abstract disposition depends on the intellect strictly defined by scientific theory. Scientific theory must be acknowledged as the spirit of art.

The æsthetic disposition depends on sentiment. It is the object of chief appreciation which depends upon tact and artistic imagination; it is particularly the collective genius of mankind and the special genius of individuals. It is obtained by nicely balancing the particularities of form, size, relation and proportion. These can be regulated and classed according to their relative importance, bearing in mind the essentials which contribute to harmonize the whole. The mind supplies these essentials, but a special vocation is required for the combination of artistical sense and practical art; yet the best disciplined logic can never supply the free active genius of the artisan.

The idea of order is essentially independent of the idea of form in an abstract sense, yet the two ideas must be co-existent in the mind when geometric solids, as well as surfaces and figured dispositions, are under consideration or manipulation.

The idea of order is the stronger when applied to arrangements of linear and successive repetition.

22. The abstract idea of order is perhaps more general than any other and seems ever present in the human mind. It refers more especially to time and distance, being associated with the sensible impression of number and combination.

Considering objects individually, we will conceive them as grouped 2 by 2, 3 by 3, &c., &c., for forming other complex objects, these being likewise re-grouped into other systems, &c., &c.

The abstract and independent combinations of ideas of order and form are, in the æsthetic sense, absolute combinations; but if we form the idea of pure combination of order or situation, we obtain a classification of ideas, not only of the assorted elements, but also of the order in which they are associated.

23. The very elements can be repeated in combination, as easily as the letters can in the alphabet, giving figures in the numerical as well as motives in the ornamental, together with multiplied figures and motives.

Permutation is an operation by which things of the same class can be arranged one with the other. Such are the rational foundations of the syntactic or science of order, i.e. exclusively logical and mathematical developments.

24. Order is Form in discontinuation.

Form is Order in continuation.

The idea of order intervenes for architectonic determination to appear in reality on the combination and arrangement of materials. This is the abstract or artificial bond which binds discontinuation and determines groups or architectonic compositions. This relation is more or less essential or homogeneous, accidental, or heterogeneous, according to the degree of independence, of unity, or individuality of each of the terms.

The determination or exact combination does not take away the beauty of idea though logically and mathematically formed ; on the contrary, exact combinations become soft and moving by the express intervention of the æsthetical causes of variation, and are made the base of disposed form and ornament. Combination is as essential in material as it is in the individual. By combination of material, high buildings are reared, and a multiplication of edifices makes a city ; man populates, and combination enacts laws, elects rulers, and thus the town and its occupants are an example of combinative power : untie the knot or disintegrate the atoms, and the individuals stand apart, lacking order as a result of a lack of combination.

The superior idea of order, which is the spirit of artistic humanity, has not yet been drawn up methodically. To logical and mathematical science we will add the architectural; these are not only positive but rational and æsthetical.

Following the method of an eminent philosopher, M. Cournot, we will classify as follows :

ORDER AND FORM.

Order purely intelligible. Phenomenal order. Logic. Mathematical Science. Architectural.

(Materials of Institution. (Signs of Convention.	Signs of Institution.
Language. | Algorithmy.	Draught. | Drawing.

Logic, mathematical and architectural metaphysics have for their object the pure ideas of abstract nature. Æsthetics have for their object the sensible ideas of human nature. Each and all these sciences have, for their general organ, language; indeed language in a general and abstract sense is the principal instrument employed in conveying logical expressions. Mathematical science has for its object, the theory of numbers and their application to the theory of sizes; its particular instrument is algorithmy. Draught has for its object the properties of extent, while drawing treats of architectural science.

If we turn back to the last table, we shall see that architectural science employs, in different degrees, drawing, draught, algorithmy and logic. Arithmetical geometry employs draught, algorithmy and logic, only. All these numerated sciences have multiplied proportions common to them. Archi-

tectural science is so complicated, so many ideas
intervene, that only the practical man can over-
come its complexity.

25. Real Form is characterized, in its individuality, by shades immediately perceived. We
will treat these shades as so many units ; they can
be grouped systematically, or used in logically constructing other forms.

If the codification of these objects were scientifically realizable, they could be theoretically defined
and classed with other scientific and positive
theories, susceptible of perfection.

All that belongs to art would, then, form a
systematic and regular doctrine, which would pre-
side over the whole system of Forms. It would
substitute a uniform method of combinations and
deductions for the occult power of invention, which
has added so materially to the world of thought and
enhanced the particular and collective genius of the
artisan, giving him a false position as viewed from
the standpoint of pure art.

If my plain speaking, in defining mechanical and
comparing it with the natural productions of the
artist, has been thought somewhat egotistical, I must
plead my excuse in a desire to show, that while there
is a certain combinative principle, there is never-
theless such variance that the two must be considered as having entirely separate existences. As
we proceed it will be seen that I shall bring both
art and science as proof of the assertions made at
the commencement of this book ; also, as we go on
we may consider ourselves as being doomed to
follow a regular and systematic line of thought, and

to give explicit obedience to certain laws, which tend to curtail freedom and subject us to the dictates of sentiment. The tendency of art is to escape from a categorically pre-conceived programme, as the result of a want of formulæ to define, and geometrical method to limit it.

26. One man may remain primitive, taking naturally what of nature presents itself without questioning the giver or the manner of bestowing ; he is content to see himself surrounded by nature and his desire reaches no further than to be able to distinguish, by name, one thing from another; logic and mathematics are no part of his existence. Another man, from his earliest being, calls for assistance in primitive distinction, yet, not being satisfied, an inborn inquisitiveness prompts him to seek the why and the wherefore : here logic plants her foot, followed by mathematical reasoning, to be perfected by a desire for a reproduction, &c., &c. Here follow the arts and sciences.

As the tilling of the soil presents itself as the first duty or man's first natural employment, so does formations upon that surface occupy the mind naturally born to become nature's student. Architecture is one of man's first studies, and architectural science cannot be pursued to efficiency without calling in the aid of kindred arts.

As one science is, in a great measure, governed by or dependent on others, so is architectural art in alliance with logic and mathematical science ; they are necessary to its solidarity. The fundamental ideas of the three sciences mentioned are the same, though essentially different in their results according to the groove one choses for his special employment.

The Aristotelian logic, taught in our schools, is the highest order of philosophical criticism, and that alone produces æsthetic architectonic science in true perfection.

We will go back to M. Cournot's theory for all that concerns philosophical developments of logic and the sciences, but for what concerns the architectural we must confine ourselves more particularly to study the past with the present. We will choose a study familiar to so many, as an illustration.

Music. The first elements in music are sounds. Sound, considered as homogeneous sizes, is subjected to variations, and these variations are the results of vibration. Vibration, reduced to numbers, is the mathematical material of acoustics as a branch of physics, these being all which intrude into this science. Sounds are, with the æsthetical particularities which follow them, the real materials of musical art. When the instrument, as a result of handicraft, is considered in connection with acoustics, mathematical and experimental science claims a place in such construction, the combination of the three being necessary to produce the result as an unit.

Logic, mathematics and architecture may, to the philosopher, be three sciences of co-equal importance and equally important one to the other, which is but natural, seeing that the philosopher's art is an abstract one, weighing and judging the proceedings of humanity. But the artist, who has personal experiences, only finds, in his inborn artistic nature, precepts, indications and prescriptions which guide him to the completion of his work.

27. The idea of order and form has been explained in the preceding chapters, yet it will be taken up again as we proceed.

28. The idea of order and the notion of form once well understood by the help of tables and signs, notations and figures, will remain a fixed element in the mind, forming a source of supply when necessary as we pass on to other categories of art work. Colour and relief are necessary elements in a work of art.

The idea of relief is correlative to the idea of form, but the idea of order is foreign. All the particularities of forming and modelling adhere to and command each other without having an abstract and independent bond. The idea of size and quantity, according to the æsthetic sense, mingles with the idea of proportion and measure.

A work of art contains the idea of proportion which exists between the projections and the reliefs of the same form. The particularities of relief and modelling must not be considered as units or as different elements to be grouped or combined systematically, as is the case with different forms and colours. Again, the power of modelling or sculpturing correlatively with the faculty to conceive or imagine space, belongs to certain privileged races. The Greeks possessed this faculty to a great extent, while the Semitic races were utterly lacking in it.

29. The sensation of colour is purely effective ;
it has not in itself any representative virtue.
Sensation distinguishes colours and gives them
specific names as a result of its distinct character
and special qualification.   Exact demonstration in
small numbers serves for comparison, to which it
adapts desinences, a mode of flexion, position and
syntax, by which means we can express, in language,
all the best varieties of decoration.

When we look upon colour in nature, or when
we conceive the philosopher's optical phenomena,
we raise scientific questioning of a theory con-
cerned with the solar spectrum, which furnishes
images always the same in themselves.   In all
other types the elements are chosen and compared
in an arbitrary manner, seeing that there is no
other bond that unites them but the frame of
human invention.

Out of this property of form that pervades the
solar spectrum, and which is produced in a
manner always identical, theories arise to measure the number of vibrations peculiar to each
colour, and count and measure by refraction that
which is produced by those colours.   But all this
scaffolding of science, strong though it seems,
does not show how the accident of colour is
bound up with the swiftness and vibrations of
ether ; or in what proportion it exists in every
luminous ray.   We will no more introduce the
solar spectrum in this work, but we will and must
use the colours generated from that system, as an
all-important element in art, i.e., the colouring or
coloured materials, but not the purely figurative,

phenomenal images, which may be considered as simple appearances only. We must accept the colours as they are in their substantial and tangible reality, with all their cortège of sensible and phenomenal particularities, and use them without putting them under scientific analysis—unimportant here.

The theory of colour is immediately bound up with the notion of form. The extent of colour and the form of separate delineation, determine the sensations of harmony and contrast of colour which would otherwise depend upon physical and physiological theory, useful and interesting in itself, but not necessary to proof in the course of reality. The notion of form is not only inseparable from decorative notion, but it contains substantial and material qualities—colouring and coloured properties.

30. The notion of colour is bound up in the notion of form and the two are bound to the idea of order. Let us take, for example, a palette ; here we have distinct elements in the abstract consideration of order; here, a number of colours, which establish variety in combination, two by two, three by three, and these again present themselves under our powers of sensorial impression.

We may agree that relief is necessary to order and form, and also a necessity in order, form and colour, for nicely determining and specifying.

The particulars of colour are subordinate to order and form, as are also the particularities of relief to form.

CHAPTER III.

MONUMENTAL AND DECORATIVE ART.

31. Ornament is constituted by a harmony of conditions, which can be shown in four parts.

Relief and colour form the basis of Decorative Ornamentation, when the relief and colours are fixed by form, and are distributed strictly according to the idea of order.

The idea of order is specialized in ornamentation and becomes a disposition.

The idea of form is specialized in ornamentation and becomes a motive.

32. Decorative art is determined in three parts, viz., Form, Relief and Colour.

The essential foundations of decorative art are Relief and Colour, which are stronger than Form.

Decorative art is an extension of decoration, which, at first, belongs to diapered forms, and has taken up a strong position wherever architectural art has become developed. An ornament is more particularly decorative where colour and relief are a lesser part than the form, surpassing it in luxuriance only.

Diapered forms which depend upon decorative art are the modelled and carved and the assembled and weaved forms.

33. Monumental art is governed by conditions of Form and Order.

Of monumental art, diapered, modelled and carved, constructed and superposed forms, the monumental stands highest. Modelled and carved forms, in conjunction with diapered forms, constitute the ornamental.

Therefore modelled and coloured decorations, with other modes, constitute, only in combination, the ornamental. Order has a special growth in ornamental form, and, as such, is an all-important factor of the whole. Order, as a growth in ornament, proves itself above logic; the one becomes a reality while the other remains as existing only in conceptive form, being, however, essential to true harmony and artistic ornamentation.

Part II.

DEFINITION AND ANALYTICAL MATERIALS.

This part is divided into five sections, treating of—(1) fundamental notions; (2) linear extent and delineation; (3) superficial extent and plane surfaces; (4) extent and corporeal forms; and (5) regularity and symmetry.

FIRST SECTION.

FUNDAMENTAL NOTIONS.

CHAPTER I.

NUMBER.

34. There are two fundamental acceptations of the word number.

1. Concrete, which is bound up with objects.

2. Abstract, which is conceptive and applied to objects in a logical form, susceptible of different uses in the reunion of groups.

35. Theory touches the very essence of number, and must therefore be treated as having three important applications.

1. Its employment for designing quantities from cardinal numbers.

2. Its employment to measure and continue sizes and to form correct ideas of quantity and measure.

3. Its employment to indicate or fix order from cardinal numbers.

The idea of number is logic employed to determine numerical and conventional signs therefore ; it must be considered as an alphabetical idea of abstract numeration, an æsthetical contemplation of things. It might be called a mental guess, under the ideas of unity, plurality, simplicity, &c. I will give the term " Indetermination " to the abstract idea of number, for by this term alone is it expressive of actual signs.

36. Number known, ideas follow ; united objects existing under different numbers, give the relation of numbers. Under Arithmetic and Mathematics the relations are determined, and they have or bear the name of proportion, such, for example, being fractional numbers. But, under æsthetic view, they are distinguished in three kinds of relation, viz., *multiple, determined* and *undetermined.* This idea of numerical relation pertains to concrete notion of disposition ; it is the one which characterizes rhythm in ornamental disposition, the composition in ornamental forms, and ordinance in monumental art. In multiple relations, the number of a series is sometimes twice, sometimes three or four times the number. In the determined relations, the numbers are fixed and determined, but have no proportion between them. In undetermined relations the particular numbers are not certain or correctly discernible, and a vague eurhythmy presides as a balancing power in particular numbers.

The idea of order united with that of quantity,

D

forming a unity, gives the æsthetic notion of rhythm, which has remarkable interpretations in the musical and poetic arts, as also in monumental art. After rhythm, which is special, comes accord, which rules the co-existence of the different parts, ordered or disordered, and blends them in superior unity. It is eurhythmy or accord which rules the ordinance of an edifice or poem and forms the particular rhythms and constitution of the work.

CHAPTER II.

SIZE, QUANTITY.

37. Mathematically considered, the idea of size is an idea of homogene, susceptible of being divided by thought into such divisions that we should like to have similar or identical parts : the divided portions may grow indefinitely. With the notion of size springs that of measure. When size is known and determined, it may be taken as a term of comparison or unity. Size thus expressed in number has the name of quantity.

Size also determines form and extent. In Architecture, mathematical quantities do not operate so much as æsthetic quantity. Figures and forms are turned into one changeable and movable unity, which is intermixed with the specific graduations corresponding with the extent of man's natural grasp of the nature of objects. From this we may distinguish two fundamental acceptations of the word size, viz., Mathematical and Architectural. A third acceptation is found, æsthetically, mixed with the architectural size ; this is also mixed, geometrically, with mathematical size.

38. Mathematical, architectural, and æsthetic acceptations of the word quantity correspond to or with the three acceptations given of size. These ideas given of size, measure, unity and quantity, constitute the basis of the science of sizes, which is an important part of the general system of Mathematics. Architecturally considered, each quantity or quality embodies its own measure, and forms unity as producing the one impression of unity. Adopt the test of mensuration, and you will find unity in size, and that number takes the place of extent.

39. Mathematical quantity is, then, determined and fixed, and is reduced, in regular method, by numbers. Architectural quantity is, on the contrary, the object of particular determinations which suffer variation following the quantity.

CHAPTER III.

EXTENT—CO-ORDINATED.

40. Geometricians distinguish two kinds of quantities, the one permanent, which is called *extent ;* the other successive, which is called duration.

All that which refers to extent is the object of immediate intuition, and the notions which result from it participate in the firmness which is essential to it. All that concerns duration is the object of perceptions whose intensity is changeable ; those elements and their co-ordination being engraved on

the mind, we can henceforth recognize them instinctively and finally as necessary to the creation
of works artistic.

41. The notions of extent and duration lead us to
the idea of space and time, whose forms rule all that
happens in the spheres of phenomenal and sensible
things. The notions of extent and duration implicate the idea of continuity which is suggested by
the contemplation of form and extent of movements
in space. The idea of continuity, which is inherent,
leads us, by logical necessity, to the institution of
mathematical science, and, by an æsthetical necessity,
to the institution of architectural science. The one
is built on the idea of homogeneous discontinuity,
the other upon the artisan's particular genius and
the collective genius of mankind in general.

Space has three dimensions, time has but one.
The three dimensions in extent possess changeable
quantities, but all that which denotes duration has
but one dimension. Following the proportion of
three dimensions, we arrive at the three fundamental
determinations, viz., quantity, size and extent, or
linear quantity, superficial quantity and corporal
quantity.

42. We conceive, in the idea of mathematical continuity, that linear extent and lines are involved
in Linear Element; that superficial extent and the
surfaces are involved in the superficial element, and
that corporal extent and space are involved in the
solid element. These conceptions are, however,
exclusively mathematical.

These elements may be said to accent works of
art. Real and corporal forms exist in extent, but

have no important place other than that they harmonize the condition of specific co-ordinations, as essential to the ideal nature of form, and conformable to the æsthetical co-ordinations of extent. All directions, orientations and oblique positions are composed of fundamental co-ordinations, and implicate them, even absolute as they appear.

CHAPTER IV.

MEASURE AND GRADUATION.

43. The idea of order and quantity is what determines rhythm. The idea of order, with the idea of quantity, determines proportion. In the first case, discontinuity is real. In the second, the ideas of quotiety and quantity are sometimes separate and sometimes assembled; for example, we do not count corn-grains but measure them.

Measure may, then, be looked upon as an unity, and is called the unity of measure. Quantity is, however, the proportion of unity of measure. The proportion of sizes is measure, and the proportion of numbers is rhythm. The complicated proportions may be fixed in three measures, thus, the subject is round when its dominant is circular; long, when its dominant is linear; oblong, when its dominant is oval.

In the prosodic quantity, it is necessary to account for the very nature of the subject, which has, for an enveloped form, a circle or an oval; for the position or orientation of the subject, the

vertical or horizontal. It is also necessary to
account for the real size, or graduation, which
modifies the accentuation of forms.

44. Graduation, or the notion of graduation,
embodies the æsthetic measure of size, and is of
changeable extent, comprised between limits upon
and under. It is necessary to distinguish three
distinct and fundamental graduations, viz., personal,
monumental and ornamental.

The personal is the normal graduation which
extends to the size of man; the monumental is
adapted to the exigences of locomotion and cir-
culation, and depends principally upon man's eye.

Normal graduation is somewhat abstract, it is a
contraction of the object to within mental sight.

45. The ideas of measure and graduation regu-
late the expansion and development of form and
ornament, necessary for detail, and to form con-
cordance in general harmony. This metric
rigorously and geometrically engenders the mono-
tony which characterizes the academic art of the
past century. Since artists have left this rigidity
and called to their assistance tools, as for example
the ruler and compass—these constitute a machine
similar to the successive parts of a drawing or
composition governed by geometrical lines—faith
has been slower, feeling a secret despair. We
should, then, hold fast to the ideas that seem rapidly
becoming extinct.

Section II.

LINEAR EXTENT AND DELI-NEATION.

46. The fundamental specifications of linear extent are straight and bent lines. The essential quality of straight lines is rectitude. The essential quality of bent lines is curvity. The fundamental forms of linear extent are, the *straight*, the *circumference line*, the *rolling*, the *spiral*, the *voluble spiral*. The *straight* is indefinite; the *circumference* is linearly finished but still indefinite in its essential quality—the curvity. The rolling is ideally indefinite, but physically finished. The spiral, coming from the straight, and forming the circumference, is indefinite. *The voluble spirals*, coming from the spiral and rolling, are ideally indefinite or physically finished. From this we see that all actual lines coming from these fundamental forms are segments, or finished lines. The fundamental determination or the segments of ideal lines are, *for the straight;* some straight, changeable in size. *For the circumference;* some arcs, changeable in size and curvity. *For the rolling;* some bent, changeable in size, curvity and declination. *For the spiral;* some spirals, changeable in size, circular curvity, and spiral curvity. *For the voluble spirals;* some spirals, changeable in size, circular curvity, spiral curvity and declined curvity.

CHAPTER I.

STRAIGHT LINE AND RECTITUDE.

47. The perfect quality of a straight line is rectitude. A straight line is the abstract image of linear continuity and only changeable as to its size. The straight is still the unity of direction, it indicates order and situation.

To form a correct idea of a straight line, we must distinguish the straight edge which limits spaces and forms.

An ideal line of rectitude is absolute, as being founded upon reason.

In edge-lines, rectitude is relative, the impression being changeable as belonging to the phenomenal world.

In the stripe line, the figuration is relative, the impression being absolute for the kind of materiality; this gives to this line an apparent and sensible breadth, which is, necessarily, imposed. Rectitude may have a tendency to pre-occupy the mind, we may instinctively take the image at first sight, which would be pretext in reality, and, in its dispositions, totality. We get from either or both sides the idea of a straight line or æsthetical idea of rectitude. The first is absolute; the second is changeable. Motion is found under most circumstances, and this is particularly essential to artistic work.

48. The metaphysical consideration of the straight, that pretends to add transcendant and sovereign qualities to the abstract idea, confers by that means an operative virtue from which, where the notion of rectitude exists, comes all the apparent and actual qualities of artistic work.

CHAPTER II.

CURVITY AND CURVED LINES.

49. It is necessary to distinguish two things in curved lines, viz., the line or draught, which is mixed with the linear element, and that remarkable quality of curvity. Curvity has the quality of continuation, and it requires a mathematical expression —size. In architecture, curvity is a particular quality which changes; yet, still, architectural quantity of the curvity can be uniform or variated from one point to another, or un-uniformly variated from the starting to the arriving point.

A line possesses uniform curvity in all its parts and belongs to circumference, and so, in variation, it is one with the bend or the rolling according to the degree of intensity of variation. A curved line, coming from the left side, can become more and more variated till the rolling is most accentuated.

If a curved line be divided into segments, curvity is more sensibly impressed; with necessarily small curves, it appears in progressive, modified form, and varies in the direction of each segment. This small curviture can be seen if the segment be denoted by a regular accumulation of points.

Curvity, in a circumference, is uniform, each of its segments forming an equal radius from a centre, and in this case the curve covers a strictly uniform line of points.

50. We have a distinct impression of curves in every point of a curved line, rolling in any way, but we mentally associate them in one curved and interminable line—the circumference.

Each of the segments of a curvity has linearly a direction determined according to the next segment and corresponding with the entire line.

Curvity being subdivided into segments, and each one being uniform in the circumference, rays, from the centre to the points in each segment, will be equal, and their being so proves the uniform nature of each segment in the circumference. By the angle formed at the centre, by any two rays, or radii, can be determined the extent of the curve. As in the circumference of a circle all lines from that circumference to the common centre are equal, therefore, any two of those lines may enclose a segment, small or large, yet uniform with the regular rolling line forming that circumference.

51. After a circle, the simplest curve is the spiral, which is uniform in its composition. The spiral has a double curved line which necessarily embraces the three dimensions of space. Straight, circumference and spirals are clear and definite ideas, upon which we can base our conceptions of curvity and curved lines.

We will now pass on from these fundamental and uniform lines to the equally fundamental but variated lines ; from the straight, and the arc, to the bent, and from the bent to the rolling lines; from the spiral to the voluble and declined spirals, i.e. to spiral rollings in space.

52. Lines are formed by the combination of two fundamental movements, the rectilineal or translatory, and the circular or rotatory movements.

1. The rectilineal movement is that made by a ruler and can only be a straight line.

2. The circular, or that line made by a compass and which must, necessarily, be circular.

3. Two rectilineal movements make a number of lines.

4. Circular, rectilineal movements make bent lines, spirals, rolling and declined spirals, and a number of geometrical lines, as spirals, conchoid, cycloid, &c., &c.

5. Conic sections can be made by combined rectilineal and circular movements with determined lengths and fixed points.

6. Two circular movements make an epicycloid or trochoid, and all curled curvities resembling starred polygons.

53. We will now treat of Descriptive Geometry and of the description of forms, i.e. the specification of different particularities which characterize them, holding each other by an abstract bond, contained in its construction. There are three kinds of affection for the linear form.

1. The analytical elements, or those which refer to the nature of lines.

2. The figurative elements, or those which refer to variated lines.

3. The æsthetic affection, viz., those introducing accent and agreement, and which are subordinate to the general form of lines.

These peculiarities hold together as a part of the solid construction, and are expressed by real form, and constitute a product of art.

54. Lines are, in their fundamental specification, unlimited. We will give here some examples of segments of straight lines, arcs and curvities.

1. Straight lines are changeable only in size.

2. Arcs, of uniform curvity, are changeable both in size and curvity.

All arcs of uniform curvity form parts of a circumference. All arcs coincide when their radii are equal. The size of an arc is dependent upon two particulars, linear extent and curvity.

As an abstract figure, a circle is always a circle, the absolute size counting for nothing but a subject, i.e. real form, and is subordinate to graduation.

All that has been said of circles can be applied to arcs and, indeed, to all curvity. In an arc the curvity can be short or long. When short, it is accentuated ; when long, it is feeble.

3. Curved lines. The curvity of a line being varied, and a product of a straight line becoming more accentuated by rolling, points out or indicates a variety of solid lines of motion.

55. There are three distinct parts in a curved line.

1. The extremity, or starting-point, where the curvity is least.

2. The extremity where the curvity is greatest, which we will call the summit.

3. The intermediate part, where the medium curvity of the extremities prevails.

When the curved line is only on one extremity it is simple; it is varied when the curve takes place at both extremities; it is prevailing when between the extremities.

The curved line has the tangent placed on the non-curved extremity. The tangent is made longer on each side, when the curvity is varied, then the tangible point separates the two constituted curved

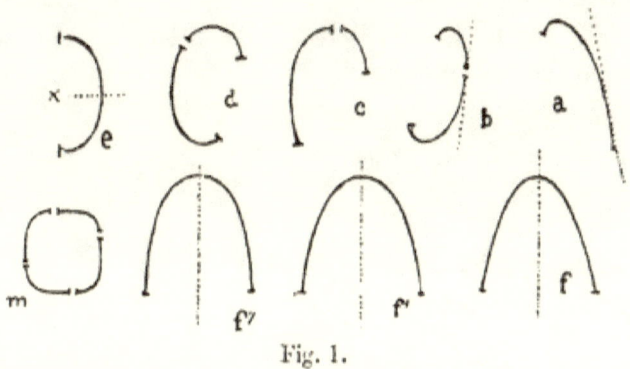

Fig. 1.

lines. The curvity is nothing between the two curved lines arising from a straight line.

56. Curved lines containing linear direction and curvity may be classed as follows :—

1. Simple curved lines, uneven and unsymmetrical (*a*).

2. Varied curved lines, even, symmetrical or ramping arcs, are of three kinds or forms. Those joined and continued by two starting-points (*b*). Those joined and continued by two summits (*c*).

Those joined and continued by a starting-point and a summit (*d*).

3. Those even and symmetric curved lines called the handles, which are of two kinds or forms : the curved lines joined (Fig. 1) and continued by the summits $f'f''f'''$; and those continued lines joined by starting-points (*c*). They follow the accent of the primitive curved line. The handles or curved lines being continued and joined by the summits give rise to three kinds : the first is quite rectilinear, with a small circularity, or the hyperbolical or angular handle (*f*), the branches being undefined.

The second is curvilinear, with a small rectitude, or parabolical oval handle (f'').

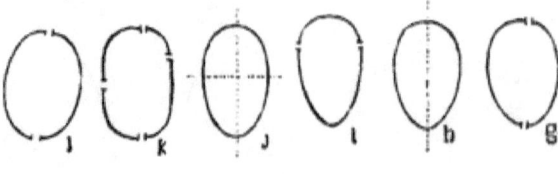

Fig. 2.

The third one, which arises from the first and second, has the elliptical or rounded handle (f''). The branches are oval and ended.

4. The oval or closed curved lines (Fig. 2) which are even and symmetrical when the two curved lines are different but directed the same way (*g*), or even and symmetric when they are constructed by the turning of the constitutive varied curved line (*h*). This regular form of oval can be made by joining two curved lines or handles (*i*). Those

two handles being different give rise to a number
of ovals, and establish continued transition of the
oval (*j*).

5. Closed or curved ovals can be even or uneven
or quartered. Ovals are even when the four curved
lines forming them are different (*k*). This form is
intermediate between the regular and the irregular
oval. The even oval is determined by the diagonal
turning of a varied curved line; the symmetry there
is diagonal (*l*). When the oval is quartered it presents the regular, typical form, and is then composed
by the turning and joining of an even symmetrical
handle, or by the quartered turning of a simple
curved line (*j*).

The form (*m*) (Fig. 1) is composed regularly by
the repetition and joining of a curved line, whose
symmetry is diagonal, preserving the transition from
the oval to the circle.

57. The draught of a curved line is quite independent of the construction of open or closed curved
lines.

58. A curved line, whose curvity is varied by
rolling round a real or imaginary point, produces a
rolling ; and this rolling can also be conceived as a
backward draught, i.e. starting from a real or
imaginary point. Here we conceive a line growing
up linearly at the same time that it forms a rolling
round this point, in such a way that, while the line is
developed, the bend suffers variation, decreasing and
progressive.

The most simple idea that can be gained of such
a graduation is, to imagine that a straight line turns
round a point, incessantly moving, while the moving

point on that line is also incessantly moving. Graduation established between those two movements, determines a rolling, whose points are rigorously fixed by the system of polar co-ordinations.

Of these two kinds of rolling are mathematical lines which constitute spirals.

The rolling considered as a form is independent of all kinds of mathematical ideas, nor can we get from geometry perfect types of that form, nor is such definition necessary to the assistance of correct idea. It is much better to draw, carve, or model at first, the volute without employing these artificial means, which are calculated to damage by reducing artistic skill to mechanical labour.

It is to be hoped that these critical reflections will be noticed, as here there is danger of the geometrician stepping in and attempting to establish rule over the artist's labour.

Æsthetic reasoning and artistic instinct are, in art, real; geometry can only be applied as an abstract science, to be verified by practical geometry.

CHAPTER III.

FIGURATIVE ELEMENTS.

59. Declination, even if complicated, embodies some of the following affections.

The joinings and inflections.

The angulations and turnings back.

The bucklings and crossings.

The branches and tactions.

If in such determined delineation we notice all these remarkable points, the parts of lines situated between them do not contain other variations than those relative to the bend; these are the straight and the bent lines, and the arcs. If we isolate all these linear segments of delineation, we break continuity, and produce discontiguous delineation.

We must, then, consider it of importance that we form a correct idea of continuity and discontinuity, of contiguity and discontiguity.

Linear continuity is maintained by joining in-

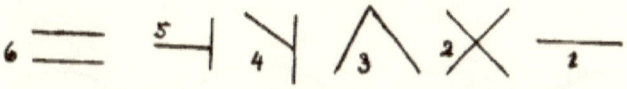

Fig. 3.

flections and bucklings. The angular branches and tactions break the continuity but preserve contiguity. Tactions and crossings are the result of discontiguity.

Conjugated straight lines. Two lines can be either crossed or cut, which determines a crossing-point (1 and 2) or contiguity, one touching the other at their extremities. This determines linear angulation and a summit (3); or, like a branch, one line reaching the other in its course, being contiguous by one of its extremities, which produces particular and complete angulation, that is, when the two lines meet together and form a right angle (5). When the two lines are separate, they have a parallel conjunction (6).

E

Two conjugated arcs. Two conjugated arcs following their reciprocal positions, determine—

1. A joining-point, when the arcs follow each other yet forming a break or curvity between them; though these lines may join, a long line may be imagined that would contain both of them, Fig. 4 (1).

2. An inflection occurs when the two arcs are joined linearly, having their cavities directed backwards or directed from concavity to convexity, and reciprocally. At the point of inflection (2) an inversion is formed in the cavity.

3. The extremities of two contiguous arcs deter-

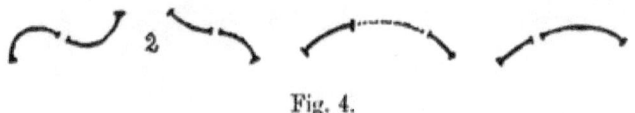

Fig. 4.

mine curvelined angulations, at the point of contiguity, or at the angular point. These angulations are of three kinds, corresponding to the three fundamental conjugations.

The first one, Fig. 5 (1), where the two arcs have their convexities set back to back; the second, where the two arcs have their concavities set face to face (2); the third, where the two arcs are turned the same way. If the arcs are equal, they produce thus two even or symmetrical angulations. The separation of the arcs is quite determined, and, if we conceive these two movable arcs round the angular point, a fixed form is apparent between all; as (1) and (2) the joining of the two arcs, and (3), an idea of their inflection. The angles (1) and (2) possess an axis of symmetry, or a linear diameter (2′); (3) has a diameter, but it is curvilinear, caused by the junction of the normal line (3′), by the centres of a series of parallels (3″), or by the centres of a series of oblique lines (3‴). If instead of looking at the angulation, we follow the course of the line, we get, in each angular point, a turning-point.

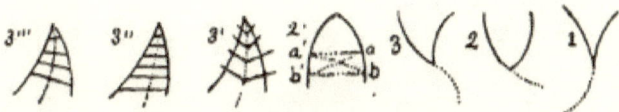

Fig. 5.

4. Contact or taction (Fig. 6). Two arcs, one touching the other at their extremities, can be equal (2) or unequal (1). If they are equal, they give a particular figure (2), symmetrical or quartered, when the arcs touch one another at their centres of figure. We can conceive that these two figures (1) and (2) are determined by the crossing of the two arcs, inflected at the point of contact, or by the joining of two appointed angulations. Two arcs, being unequal, give rise to a particular form of contact, which is inside (3), the arcs being directed the same way. This particular form may be considered as a result of the intersection of two arcs, and where one cuts the other at the point of taction.

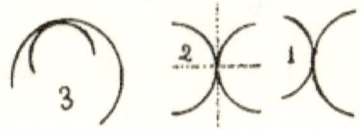

Fig. 6.

E 2

5. Crossing two arcs (Fig. 7), where one cuts the other in a point, determines a crossing-point (1) and three kinds of curvilinear angulations (1′) (1″) (1‴).

Two different or equal arcs (Fig. 6) can also be cut in two points when the convexities are turned back to back, Fig. 7 (2). If the arcs are unequal and turned in the same way, they produce a figure (3).

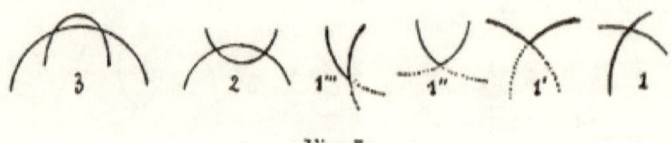

Fig. 7.

6. Branchings (Fig. 8). Two arcs meeting at extremities produce a branching (1). If the arcs join together, they form a branching-point, a joining-point, a point of inflection, and a point of contact (2). The branchings of the figure (1) contain the forms of angulation.

7. Two arcs are discontiguous, or separated (3), and are yet alike, being symmetrical or parallel.

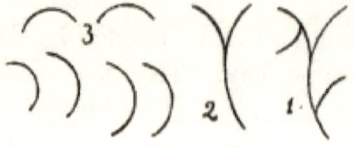

Fig. 8.

60. Two conjugated curved lines (Fig. 9).—Figures produced by curved lines are varied by the accent of curvity, although it may not widely differ from the preceding one; but starting from the curvity, two curved lines or figures appear individually. These figures are rolling or curling.

The curl determines a closed figure (Fig. 9) and a crossing-point, (1) and (2). In the case of regularity, the curl has an axis of symmetry which passes by the crossing-point (2). As rolling is a mode of figuration belonging to linear continuity, it gives rise to different forms. The course of a straight line can be stopped by a round or oval curl (3); so also the course of an axis (4).

By means of this intervention of inflection they pass from convex angulation to a curl, by two inflections, which meet at the crossing-point or curling (5), or from the crooked angle to the curl

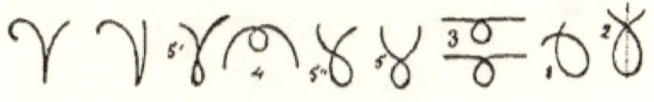

Figs. 9, 10, 11.

by one inflection (5″). They pass from the concave angle to the curl by four inflections (5′). It is necessary to distinguish both curling and intersections.

CHAPTER IV.

ÆSTHETICAL AFFECTIONS.

61. These affections are of two kinds, linear and figurative. They are linear when subordinated far from the pace of the line and introduce variety by discontinuity. They are figurative when they determine ornamental delineations in engrailing, lacing, twisting, &c.

62. For restoring the pace of a line of form (1), we determine some attractive points which follow their approaching, and interrupt the very law of form. These points, bound by straight lines, determine a polygonal line which, in the limit of our perception, is equal to a perfect line. They substitute discontinued variation for continued variation. This discontinuity is realized by the artisan in carved, turned and hammered forms. These discontinued variations are so many accents, so many marks, from which the figure is determined and which presents to the mind a real subject which reveals the handiwork of the artisan.

CHAPTER V.

DELINEATIONS.

63. There are two kinds of delineations.

1. Delineations of profile and outlines of form.

2. Simple delineations.

Profiles and outlines of form are the conventional and abstract delineations, which have no existence independent of form. The fundamental dispositions are the joinings, inflections, and angulations. The joinings are the starting-points, which represent linear construction. The inflections are not reduced to a point; they can be either short or long, so that two curved lines are abruptly drawn closer or otherwise, that a straight line may separate them. All the particulars which characterize extent and linear form are found in the simple delineations with accentuations infinitely varied, though they may arise from the technical mode of decoration. These delineations constitute the drawing of ornament and form.

In ornamentation these particulars, in their essential and proper nature, give rise to a number of dispositions, although without any accession of æsthetical ideas or abstract principles of monumental co-ordination.

SUPERFICIAL EXTENT AND PLANE SURFACES.

64. The fundamental specifications of superficial extent are plane and curved surfaces. A surface can be plane, or can be made up of plane, curved and round surfaces, having also concavities and convexities following in order with faces, straight or curved edges, polygonal or curvital.

It is necessary to understand clearly that superficial extent and surfaces are co-existent with a number of smaller superficial elements. Surfaces, like lines, are submitted to a principle of co-ordination, contraction, disposition and symmetry, which determines the individual forms.

As surfaces are naturally inherent in the material or æsthetic form, we shall notice those plane surfaces only which are referred to in this work.

CHAPTER I.

FUNDAMENTAL FORMS OF SUPERFICIAL EXTENT.

65. These forms of superficial extent are—

1. *The plan or four-cornered space.*
2. *The circle or circular space.*

3. *The label or linear space.*

4. *The angle or angular space.*

The superficial size or plane surface is homogeneous and uniform. An uniform and declined curvity is not the same as a plane surface, for a number of plane surfaces can be held within the limits of the form which contains them. This essential uniformity in superficial extent necessitates the idea of co-ordination.

1. *The plan.* The harmonical, co-ordinate lines of the plan, are the horizontal and the vertical. These correspond with the two dimensions of the plan. Oblique and diagonal lines also participate in the fundamental dimensions of the plan. Oblique lines have two fundamental positions. Horizontal, vertical and oblique lines determine four systems which cut the plan into quadrilateral segments, when combined two and two, three and three, and so on. It is this harmonical decomposition of the plan which gives a particular quality to monumental decoration.

2. *The circle.* Suppose superficial extent to be contained in a point, and then suppose this point to extend uniformly, then the correct idea of a circle is formed. The harmonical, co-ordinated lines of circular space, are those striking off from the point and cutting the infinity of concentrical circumferences which fill the whole circular space. These concentrical circumferences limit the circular space, although in itself unlimited. The rays from the centre cut the circular plan into angular spaces or sections. These two systems of co-ordinated lines, reunited, cut the circular spaces into four-cornered

and curvilinear segments, which are the segments of the wreaths, determined by the system of concentrical circumferences.

3. *The label.* Superficial extent, being reduced to a straight line, is indefinite, although limited. The co-ordinated lines of the label are those that run parallel to the head straight line, also the transverse straight. These two systems of co-ordinated lines cut the label into four-cornered segments, like those of the plan.

4. *The angle.* In angular space, the superficial extent is declined from the opening to the summit of the angle, where this extent comes to nothing. It is necessary to know, by angle, the superficial space which lies between the two oblique straight lines, or the space formed by the fold of a straight line.

66. In order to value the sizes, mathematically or architecturally, we must reduce them to unity of measure or appreciation.

The unity of measure, for valuing sizes, is the square.

The unit of appreciation, for the contemplation of sizes, is the circle.

67. In architecture, where the elements so differ, analogy is held as necessary to distinguish the two methods of appreciation.

Applying the general method of appreciation to the fundamental forms or spaces—i.e. to the label and circle only—we shall arrive at an exact definition of what delineation of a plane surface means. All circles described and united within the limits of the label are equal, Fig. 12 (1). This is an artificial

method, for we have immediate perception of linear
and uniform extent, as the unnecessary aid to the
true artist.

This is also artificial when applied to the angle,
yet it has the advantage of placing before the eye
the essential quality of the angle, that is, its delineation (2), (3). If the angle is either acute or obtuse,

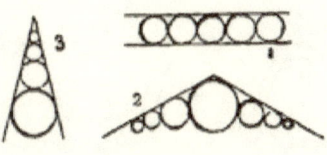

Fig. 12.

the declination of the space is continued, i.e. unilateral or collateral.

CHAPTER II.

GENERATION OF THE FUNDAMENTAL PLANE SURFACES.

68. The architectural unity of appreciation of
plane surfaces is the circle, where the extension goes
round an actual or vertical point. The extension of
size is determined by combination of circular extent and linear translation, the centre of extent
going over all the points of the headline. The headline is the head-point. These are the irreducible
and last foundations of the generation of forms.
We conclude that the correlation from the superficial
extent to the linear is absolute. A proportion established between the circular extent and the head-

line is what determines the fundamental form of the plane surfaces.

The general terms of this proportion are—

I. A fixed circle and a headline; the form of the circle is then determined by the translation of the circle, following the course of the line.

II. A changeable circle, whose size changes harmonically with the form of declination and a head-

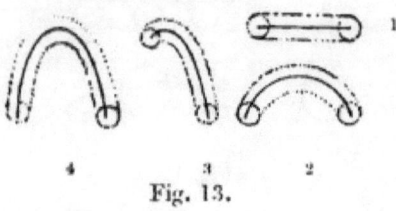

Fig. 13.

line; the form of extent is then determined by the translation of the circle in all successive points of the line.

I. *A fixed circle and a headline.*

69. Straight headline. The straight headline, being indefinite, is, by the translation of the circle or the label, indefinitely symmetrical in all its parts, and is thus uniform. This is the third fundamental form of superficial extent. If a straight line be finished, we can have a figure, equally finished, and ended at its extremities by a circular form, Fig 13 (1).

Circular headline. The circumference being closed, we get a circular label; the arcs being finished, we get some figures finished and ended at the extremities by a circular form (2), by the closing of the arcs.

Curved headline. The bent line being finished, we get the figure (3). The handle will give the figure (4).

Further explanation is unnecessary upon this subject of the mode of generation, applied to simple lines, composed or composite, i.e. to handles, ovals, ovates, undulated lines, either turned back, buckled or polygonal. Æsthetically finished plane lineary surfaces are segments of indefinite lineary plane surfaces; the full arc is a finished plane circular surface; therefore it will be seen that we must not confound physical with æsthetic terminations. The extremities of lines are physical terminations as opposed to harmonic.

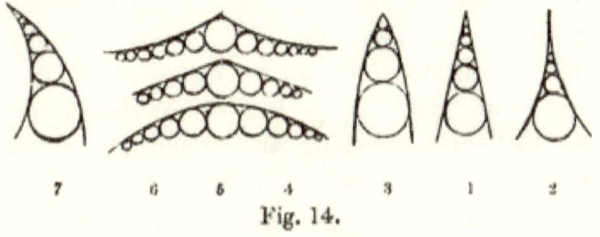

7   6   5   4   3   1   2

Fig. 14.

Linear plane surfaces are composed of rectangular segments, with æsthetic ideas of centre and circle overruling the notion of lines.

II. *A variable circle and a headline.*

70. Straight headline. Taking a fixed circle and passing its centre over every part of the headline, the decreasing circle gives a correct idea of what is termed a declined plane surface, presenting, at the same time, angular declination. Again, take a straight line as a declinator, and recline it to the headline, or part of the arc, and you have exhibited specimens of rectilineal, raised and sharp angulation, Fig. 14 (1), (2), (3).

We must form ideal circles at all points of the

declinator. We must also realize the idea of tangent
circles, which are the essential types of declination.
In Fig. 14 (1), the declination is regular, and shows
the ordinary rectilineal angle. In (2), the declination is rapid and the appointment accentuated.
In (3), the declination is slow and hesitating,
and the plane surface is stronger than the angular
declination. Order of forms should, therefore, be
1, 2, 3. In these figures the declination is prolonged, although following the inclination of the
envelope-line. The declination changes the type,
for if the angles be very open, the declination is
blunt (4), (5), (6).

This declination becomes symmetrical or recurrent, and goes by each side of the headline.
These declinations belong to the horizontal order,
conforming with the second type of a straight line
(symmetrical at every point). The first belongs to
the vertical order, and conforms with the first type
of the straight line (a successive line from one
point to another).

Following their graduation, these declinations are
forms or desinences. These desinences give detail
to form and impart degrees of accent.

71. Arched headlines. The progression of a
circle following an arc determines a particular form
of angular declination—circular or hooked (7).
From this declination we must distinguish two
fundamental forms.

1. The angular declination, where the figure
ends in nothing, and the headline going as far as
the summit of the angulation.

2. The oval declination, following the head-

line at the extremity of which, rays, a circular plane surface, goes round the desinences. In angular declinations the headline can either be a straight line or an arc.

The series of circles is indefinite on the open side of the handle, but finished near the summit, in a fixed point where the circle is osculatrix of the curvity. The oval desinence is more or less accentuated, and changes from the arc of the circle as far as the angular desinence.

Though the handle may be hyperbolic, parabolic or elliptic, and the headline curved or straight, we

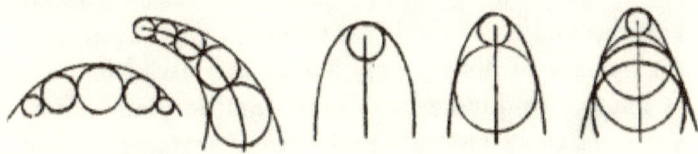

Fig. 15.

get (Fig. 15) the figures (1), (2), (3), (4). The elliptic form (3) is half an oval. The curved form (4) has an inflecting point.

With obtuse angulation, we get oval declinations, resembling the handle of a basket.

All the forms of plane figures are synthetically made up of the following elements.

1. The circular extension.

2. Linear prolongation, either uniform, recti-linear or curved.

3. Angular declination, rectilinear or curved, straight or bent.

4. The angular declination, rectilinear, straight or bent. Angular declination is physically finished,

and its plane figure is of the same form in all its parts. The oval declination is harmonically finished, and its plane figure is composed of circular extension and angular declination.

CHAPTER III.

FUNDAMENTAL DOMINANT LINES OF PLANE SURFACES.

72. Extent, being essentially uniform in all its parts, is harmonically composed of two dimensions ; all fundamental forms of extent, and segments of them, participate, at all points, in its essential nature, and depend, in a general way, upon the two principal dimensions, length and breadth.

These two dimensions determine the plan. Segments of the plan are simple plane surfaces, whose actual dimensions are determined. These segments are, generally, polygonal.

If in a plane surface one dimension overrules the other, we get a third form of superficial extent, or label. The segments of a label possess a fundamental character, viz., linear prolongation following the course of the line. This is a particular kind of plane surface, uniform in all its points, but where a dimension is reduced the other grows up slowly, and is the circular space for immediate contemplation in all the successive points of the headline. In the label, the plane surface is indefinitely symmetrical from both parts of the point under consideration.

We will imagine, by proportion to the dimension, which characterizes the label, that the space is not uniformly divided, but goes on decreasing from one extremity of the line to the other until it ends in nothing, and we get here a fourth form of superficial extent, or angular declination. All segments of this extent possess angular plane figures.

73. Spaces being theoretically undefined and of homogeneous composition, an architectural determination is fixed for imagining the existence of harmonic segmentation in these spaces. Each of these gives rise to particular and proper segments, which participate in its fundamental nature.

1. Segments of the plan. The extent of the plan being sub-divided by harmonic lines, gives rise to polygonal, four-cornered segments, the dominant line being in the plane surface.

2. Segments of the circle. The circular space being reduced, regularly or irregularly, in segments, by straight lines, polygons are produced. The dominant line is circular.

3. Segments of the label. Linear space being separated by straight lines, we obtain four-cornered segments. Their dominant line is linear.

4. Segments of angles. Angular spaces produce proper triangular segments. The dominant is angular.

Resuming the fundamental forms of extent, it is necessary to conceive the following correlations, viz. :—To rectangular space corresponds arranged repetition, or simultaneous order, correlative to the system of lines and network which bind the different points of extent.

To circular space corresponds circular order, correlative to regular repetition of points round a centre.

To linear space, or label, corresponds uniform linear order, i.e., indefinitely symmetrical form on both sides of each point.

To angular space, or angle, corresponds successive linear order, i.e., everything on one side and in the way of the line, or from one point to another.

CHAPTER IV.

FUNDAMENTAL FORMS OF PLANE SURFACES.

74. Combining the fundamental modes of generation of plane surfaces with the diagrams, we obtain

Fig. 16.

Fig. 17.

linear prolongation, and declination of the simple or compound lines which form the plan of disposition, and numberless figures which are proportional to the following fundamental forms (Figs. 16 and 17).

1. The circle. Absolute unity or fundamental bases of forms are entirely centred in a head-point; its length varies with the ray of extension, Fig. 16 (1).

2. The angular oval. Starting from the circle,

we conceive the declined fluxion of a plane surface. Following a head-line, coming from the head-point and finishing at the other extremity, following also the form of delineation, which can be rectilinear, sharp or swollen, we get the three principal forms of the angular oval (2) (3) (4).

These angular declinations form desinences for the plane surfaces. If we take a head arc, or a curved line, instead of a straight line, we get the fourth form of the angular oval, i.e., the curved or broken oval (5).

3. The oval proper. If starting from a circle we conceive the declined fluxion of the plane surface, following a head straight line, coming from the head-point and harmonically finishing at the other extremity, i.e., forming at the extremity of the head-point, from whence it arises, a new circle, we get the oval properly so called (6) (7) (8), which, at the joining of the angular oval possesses a higher character of unity, i.e., harmonized form. The unity of angular ovals consists in an actual ending of the plane surface. The circular extension of the extremities of the headline can pass, together or separately, by all the kinds of size. If the circle of desinence be small, the proper line is inclined towards the angular oval. If the headline be an arc or a curved line, we get the curved oval (9).

4. The oval. Starting from the circle, we conceive the oval fluxion as going symmetrically on each side of the head-point, following a diameter, we get the oval, Fig. 17 (1) in its quartered form, and possessing a centre of figure, and two head-points

which are symmetrical with it. If the circles of
the extremities are reduced to a point, we get the
angular oval (2).

5. The hooked line. The straight line and the
arc, being in uniform composition, are symmetrical
or even.

The curved line, being infinitely varied from one
point to another of its course, is an uneven or unsymmetrical line. If, then, we conceive a circle at
one of its extremities, and the declined fluxion of the
plane surface following the course of the line, we
shall be able to determine two distinct forms,
Fig. 17 (3) (4).

In (3) the size of the space is declining from the
starting-point to the desinence; in (4) the size of
the space is declining from the point of desinence
to the starting-point. If the circle of the extremities were reduced to a point, we should get hooked
lines, with a regular desinence. If the headline
became rolling, declination would take place, exclusively, from the starting-point to the desinence.
The Ionic style is a remarkable example of this
form.

CHAPTER V.

HARMONICAL LINES OF THE FUNDAMENTAL FORMS.

1. Harmonical lines of a circle. The circle
possesses two systems of harmonical lines, viz., the
rectilinear rays coming from the centre, and the
concentrical circumferences which limit the extension of circular space.

2. Harmonical lines of the oval. The har-

monical, co-ordinate lines of the oval are, the principal centre and a headline, i.e., the centre of extension and a prolonged line. If we wish to mark the lines of radiation, we will trace a tangent on a point of a headline, then the normal line which will cut the next oval, and, at this point, trace a new and normal line which will cut the third oval, and so on ; we shall thus get curved lines coming from the principal point, producing a palmated harmony. We have, then, the harmonical lines of the oval and circle ; some of which are curved lines

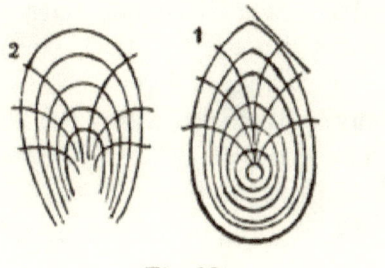 

Fig. 18.   Fig. 19.

of changeable declination, and some concentrical lines which are handles of declination of curvity— also changeable—depending harmonically on the handle-envelope of the oval, Fig. 18 (1) (2).

The form of the oval being essentially changeable, this harmonic sub-division is essentially a subject of tact and artistic feeling.

It is necessary to conceive that, starting from the circle subdivided harmonically, the head-point comes successively in all parts of the headline, carrying away with it all the lines of the primitive network which do not modify the general co-ordi-

nation of the form. The fluxion or the declination
of the parts is not thus harmonically equal at
what makes the essence of the oval, as a result
of the uniform mechanical movement, Fig. 19 (1)
(2).

But if we choose the head linear line and proportion the oval, we notice that that oval is determined
by the successive transfer of a circle of extent,
changeable in all points of the headline. The

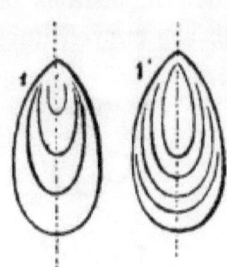

harmonical lines are the headline, or virtual, and some symmetrical branches on each side
of it. The harmonic lines are
subordinate to the centre.
The branches have their concavity turned towards the axis.
The curvity of these nerves
varies with the curvity of the
outlines, the outline being recti-

Fig. 20.

linear like the nerves.

By this second sub-division, the plane surface
(Fig. 20) is pennate with a propension at the
palmation, i.e., at the occult convergence of the
nerves on a secret point (1) (1').

3. The harmonic rays of the hooked line. The
course of a hooked line being linear, it is necessary
to conceive that the successive centres are multiplied
and separated in proportion with the curved line,
and with intervals which follow, in this variation, the
laws governing the declination of the head curved
line. Here we get (Fig. 21) the headline and
those embracing it, together with the nerves which
cut them. The nerves on the side of the cavity,

and those on the side of the concavity have their
curvity turned the same way.

75. We now pass from the consideration of the
hooked lines to that of the forms and pennate
sub-divisions which are correlative to the oval or

Fig. 21.

angular declination. In oval declination, at the
point of ovalation, stands a centre of palmation; and
following the curvity of the branches, two collateral
series of curved and rectilinear rays follow in a

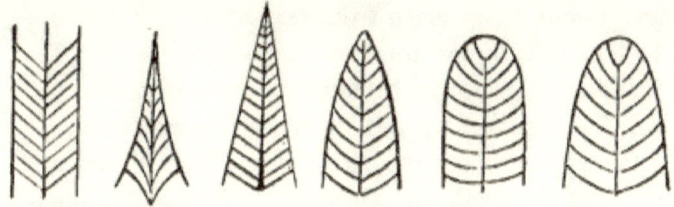

Fig. 22.

slanting direction towards the occult convergence.
In the angular and convex declination, the collateral
nerves of oval declination follow the laws of declination, except in the palmation of the summit. In
convex declination the collateral nerves follow the
laws of variation of nerves in the oval. In rectilinear declinations, the nerves are rectilinear or
curvilinear, but parallel.

From this we see that transition, which joins the
circle and label, is continued, passing by the oval to
the angular and oval declinations. These three
diagrams should be noticed:—

1. The radiated disposition which is subordinate
to the head-point.

2. The pennate disposition which is subordinate
to the head-line.

3. The palmated, which is subordinate with a
headline and point, and, therefore, is in connection
with the others mentioned.

CHAPTER VI.

VARIATION IN THE FUNDAMENTAL FORMS.

76. All forms have two dimensions of extent.
Three minor cases arise from these two, viz. :—

1. The two dimensions are equal.

2. The longitudinal dimension is the greater.

3. The transversal and longitudinal are equal.

The types corresponding to these three cases are
the circle, the oval, and the ovate. The two dimensions in a circle are equal; the shape is concen-
trated entirely in a point. In the ovate the
dominant dimension is longitudinal, and it is by this
dimension that the oval shape is susceptible of
determined orientation, straight or overturned, i.e.,
equal to vertical order, or slanting in every degree,
and then reduced to a determined disposition. In
the oval, the dominant dimension is transversal,
and equal to horizontal or symmetrical order.

A circle faintly depressed on one side, forms an

oval whose determined dimension is longitudinal and the sense vertical. A circle depressed on two of its parallel sides, becomes an oval whose dominant dimension is transverse and the sense horizontal. An oval lengthened on one side only becomes an ovate.

Angular or oval declinations, straight or bent, are forms lengthened and measured by the circle, the oval or the ovate form.

If the curved declination becomes rolling, the unity of shape would be a circle, containing a central pole or point around which turns the line, while the curved lines possess poles which exist linearly.

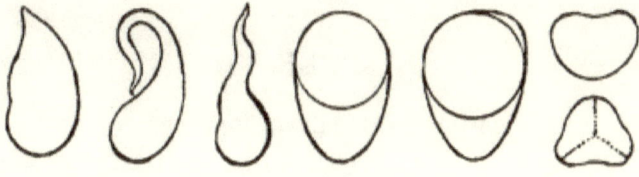

Fig. 23.

The head-point of a circle implicates a number of axes or rays crossing the point and terminating at the outline. The two head-points of an ovate possess the variations of the circle; the headline can be either straight, arched, bent, polygonal, or broken. The oval participates in both circle and ovate forms, suffering all their variations in their determined dimensions. In the diagram, the headline can be rectilinear or curvilinear, uniform or varied, &c.

77. The variations, which belong particularly to the outline, remain stationary in the, properly called,

plane surface. There are two ways to draw, viz.,
by designing all the parts of the outline by hard
lines, and by following the whole outline by an un-
interrupted line. In the first, we show the accent
without losing the integrity of shape; and, in the
second, we lengthen the continuity and neatness
of the touch. An outline may implicate a plane
surface, real or imaginary, implicit or explicit.
When the plane surface is real, it is the outline
which is implicit or explicit, and the modes of
ornamentation, following these conditions, shall
vary.

CHAPTER VII.

POLYGONAL PLANE SURFACES.

78. The essential character of polygonal plane
surfaces is that they are segments of plane surfaces, i.e., portions cut out of the superficial extent.
The number of segments which form the outline is
indefinite; and the segments, multiplied indefinitely,
pass insensibly from a rectilinear to a curvilinear
outline.

What strikes the sight are figures of touch,
or described polygons, in the rectilinear outline.
Here it is necessary to notice segments of straight
lines, which meet together end to end, forming
equal and unequal angulations at every point. According to the number, length and inclination of
these segments, together with their general co-

ordination, we get different figures. At least three straight lines must be employed to determine a closed figure. Those figures with three sides are triangular, four sides = four-cornered, and when a larger number of lines are employed we get polygons or multiple figures.

79. The idea of number correlative to the idea of different parts, and the idea of grouping correlative to the idea of harmony, give a conception of polygonal figures. We have, then, the immediate perception of polygonal figures, and the intuition of a plan of co-ordination is joined, which arises from the idea of order, which geometry renders sensible, by giving direct representation, aided by definitions of symmetry. The plan of co-ordination is the unity of figures, consequent upon harmony, or a collection or group of parts whose number can be estimated, determined and given in numbers.

In geometry, the idea of a group is given by the touch which furnishes direct representation. The sensible touches, whose particular properties give the properties of extent, pre-figure the particular and constitutive units of a group.

80. Following the series of polygonal figures, we distinguish the triangle, the four-cornered, the pentagon, the hexagon, &c., which successively correspond to the numbers 3, 4, 5, 6, 7, 8; 1 should design a touch, 2 an angle. Numbers are of different kinds, even, multiple, &c., and to these different kinds of numbers correspond particular kinds of polygonal figures; this correspondence is established by particular conditions of symmetry and regularity following the diagram of disposition, whose shape is relative to the composition of numerical groups.

I. TRIANGLES.

81. No. 3 is absolute, because it forms an irreducible group. The triangle is also an absolute and irreducible figure. The form is regular or irregular as determined by the sides being equal or unequal (Fig. 24). If they are unequal, we get the scalene triangle (1), whose particular form is rectangular, if two sides determine a perpendicular angulation (2). If two sides are equal, we get the isosceles triangle, which is of three forms.

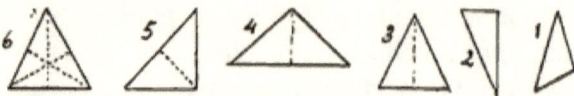

Fig. 24.

The first has an acute angulation (3), the second an open angulation (4), the third a perpendicular or absolute angulation (5); and if the sides are equal we get the trigon (6), which has the form of an absolute triangle.

The summit of a triangle being more or less distant from the base, we get the two æsthetical aspects quite different.

The raised triangle and the elliptical triangle. The triangle, being proportioned to the circular space, has not an angular denomination; it is, essentially, a regular or irregular segment of a circular space, the simplest of all, and which has three variable, angular desinences. The absolute form is the trigon, which has three equal angles and desinences. Every triangle, whose domination is a circular halo, if not absolute or trigon, is an irregular form. In the case where this form would be symmetrical, or isosceles, the impression should not be less dissonant, the type of the isosceles being angular, raised or elliptical. This form is, essentially, regular; the trigon has its angles equal as well as its sides, it is a central figure, which has no absolute orientation; it envelopes a regular straight line. The natural lines of a trigon are (Fig. 25) the heights (2), the rays (3), the apothegms (4); the

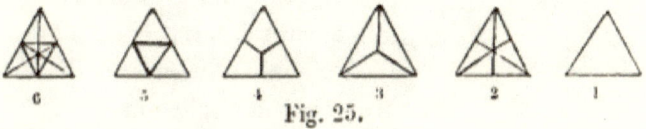

Fig. 25.

figure (5) is the inscribed trigon, and figure (6) has all these particularities.

82. If we establish a correlation between the base and the height, it is that the height is determined by the two rampings of the triangle. Every inclined line being relative, having two absolute directions, the horizontal and the vertical; and further, these two rampings and these two inclined lines being of equal inclination and length, give a view of the vertical, which is their normal, absolute line. Between the isosceles triangles, the geometrical form is the rectangular isosceles triangle. If we trace a series of isosceles triangles, whose heights and lengths are rigorously determined, starting from a base equal to the number 8, and giving the

heights successively 1, 2, 3, 4, : : : raised from that
base, we should get the following list (Fig. 26) :—

Base.	Height.	Sides.	
8	1	4·123	
8	2	4·469	
8	3	5	Pythagoras' isosceles triangle.
8	4	5·656	Rectangular isosceles triangle.
8	5	6·403	Egyptian triangle.
8	6	7·211	
8	7	8·062	

83. Applying to a few isosceles triangles of
different forms the general method of the inscription of tangent-circles (Fig. 27), we will consider
what constitutes harmony, beauty and unity in that
triangle, among others.

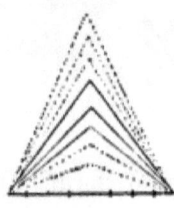

(1). Rectangular isosceles triangle—abrupt bilateral declination.

(2). Trigon—abrupt and divided
in three part declinations.

(3). Declination greater in
height.

Fig. 26.   (4) and (5). Declination more or
less accentuated in height.

(6). Declination in height quite decided, the
declination on the base disappearing.

(7), (8), (9), (10). Bilateral declination more and
more prolonged.

(11). Declination exclusively ramping and prolonged horizontally, all vertical declination disappearing.

The artifice of the inscription of circles has the
advantage of showing how a form more or less
favourably follows its graduation or actual size.
When we say that the horizontal or vertical has
disappeared, we mean that they are more or less

eclipsed. The erasure is seen in the preceding figures, but we must understand that if the figures were greater, that which remains, after the inscription of the circles, would be perceptible and should leave a shade of indecision which would render the image not clear.

If the isosceles is ramping, or on the contrary raised in only one vertical declination, we have

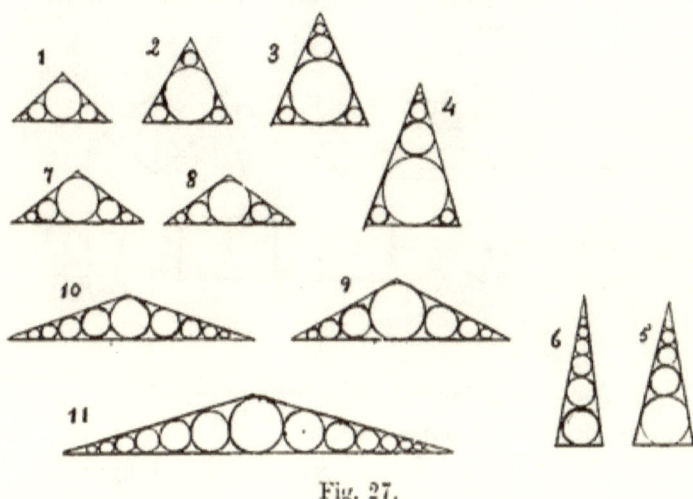

Fig. 27.

two types, and we get æsthetical determinations quite distinct and heterogeneal; for example—in architecture the pediment, and in the particulars of ornament, the obtuse or acute desinences.

II. Four-cornered Surfaces.

84. Quadrilaterals are regular or irregular segments of the four fundamental shapes of extent, the plan, label, circle and angle. If they are segments of the plan, the dominant is the plane surface circumscribed by the quadrilateral figure; if they are segments of the label, the dominant is linear; if segments of a circle, the dominant is circular; if the segments at the angle, the dominant is angular. Quadrilaterals have four sides, equal, unequal or varied, and are submitted to a plan of construction or to a diagram of symmetry which gives rise to eight kinds of forms (Fig 28).

1. The irregular (1), unsymmetrical quadrilateral, or those without an axis.

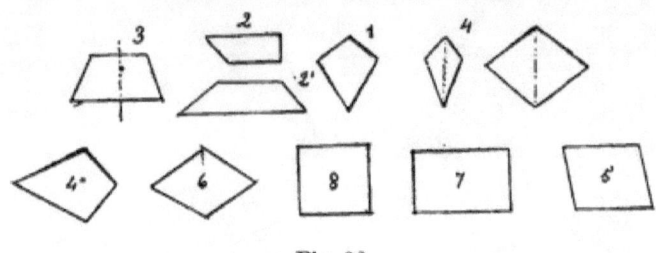

Fig. 28.

2. The irregular (2, 2'), unsymmetrical trapeziums, or those having no axis.

3. The regular (3) and symmetrical trapezium, having an axis.

4. The variable and symmetrical corners (4, 4', 4''), according to angle, with an axis.

5. The parallelogram (5), whose symmetry is diagonal, and with axes.

6. The quartered lozenges, whose symmetry is diagonal, with two axes.

7. The quartered rectangle, whose symmetry is diagonal, with two axes.

8. The square with the symmetry, four axes.

The quadrilateral has its four sides unequal and

articulated under unequal angles; the parallelogram has its sides equal 2 and 2, and parallelly articulated under equal angles 2 and 2 ; the middle form
between the quadrilateral and the parallelogram
being the trapezium, which has two parallel sides.
The two particular forms of the trapezium are the
rectangular trapezium, and the isosceles or symmetrical trapezium. The rectangle and lozenge are
particular kinds of parallelograms : the sides of the
lozenge are equal, so also are the angles 2 to 2.
The sides of the rectangle are unequal, but the
angles are equal and straight. The absolute shade

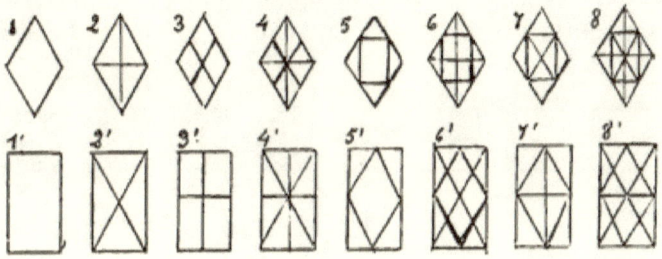

Fig. 29.

of the movable lozenge is the square, which has
its sides and angles equal. Symmetrical lozenges
have their middle shade between the quadrilateral
and the lozenge, their composition being varied
according to the angles; an acute angle, an obtuse
angle, two obtuse or two acute angles opposite the
shade of the corner, is midway between the irregular
quadrilateral and the lozenge. The above examples
show the narrow correlation which exists between
the rectangle and the lozenge. The natural lines
are the diagonals, which join the summits, and the
medium which joins the middles of the opposite sides.

These lines are axes of symmetry. In Fig. 29, (4) (4′) form a lozenge or rectangle; (5) (5′) reunite the diagonals and mediums; (6) (6′) show the inscribed figure; (7) (7′) unite the inscribed figure and the diagonals; and (8) (8′) unite or show all the peculiarities. The square participates of the rectangle, and the lozenge is the absolute form between the quadrilaterals.

85. The dominant of the square is the circle. The dominants of the rectangles, lozenges and corners are even ovals, quartered for rectangles and lozenges, and even for the corners. The dominant is even for an open space, i.e., the oval implicates the rectangle, and the rectangle implicates

Fig. 30.

the oval, or the oval dominant which determines the æsthetical appreciation of the rectangle. The oval and the rectangle each have two axes, an oval with undefined axes, and an uncertain form whose dominant is a circular space; but if the axes become plainer, the dominant becomes linear. In the two cases the forms are extreme, one condition being exaggerated to the detriment of the other, which suffers for the want of development. This co-existence of two different conditions determines the variety of ovals and rectangles. In the lozenges, angular desinences are most impressive; if these are prolonged, the oval also is prolonged, and is less adapted to the form. The corners and ovate figures have varied forms; the abside and desinence can be varied in every point, as the figure is lengthened or otherwise.

III. Pentagons.

86. Pentagons are of three kinds, regular, semi-regular and irregular (Fig. 31). The irregular pentagon has sides and angles unequal and no axis of symmetry. The semi-regular pentagon has lines and angles equal, and the figure is symmetrical in proportion to an axis.

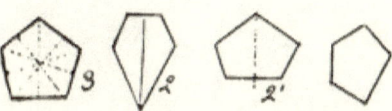

Fig. 31.

The regular pentagon (3) is a centred figure, and has for its dominant a circle, or circular arcola. The mid-regular pentagon (2, 2') is an even figure, and has for its dominant the lengthened ovate or elliptic; a regular pentagon has for its dominant a plane surface, the circle and the angular desinences.

IV. Hexagons.

87. The hexagon (Fig. 32) can be regular or absolute (1); in this case it has six axes of symmetry, three angular and three transversal; its dominant is the circle. The semi-regular hexagon has two forms (2) and (3); the first has its sides equal and three angular axes; the second has its angles equal and three transversal axes; their

dominant is the circle. The quartered hexagon is of
two forms, one rectangular, the other rhomboidal
(4) (4'); their axes are quartered, one angular, the
other transversal; and their dominant is the oval.
The diagonal hexagon has several axes, or a circle
of symmetry (5) (5'). The even hexagon has an axis
of symmetry, and an ovate dominant (6) (6') (6")
(6'"). Lastly, we have the irregular hexagon (7).

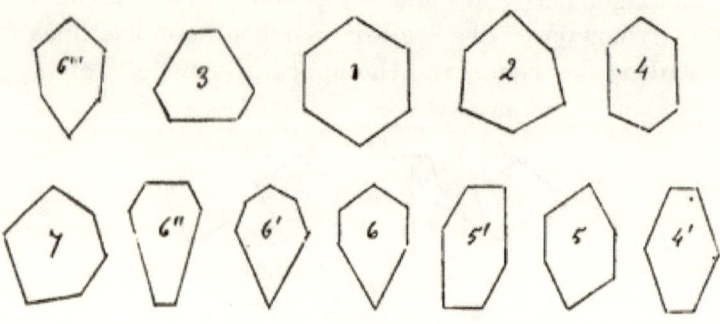

Fig. 32.

V. Octagons.

88. The number eight can be separated into multiples as, 8 = 4 × 2, or 8 = 8 × 1. We get six
distinct octagons (Fig. 33). The regular octagon
(1) has eight axes of symmetry, four angular and
four transversal, and has for its dominant the circle.
The semi-regular octagon (2) (3) is of two forms,
one having two angular and two transversal axes,
the other four angular axes; its dominant is the
circle. The quartered octagon is of two kinds, one
having a transversal axis, the other an angular axis
(4) (5); the dominant is oval. The even octagon is
of two forms, one having a transversal axis (6'), the

other an angular axis (6); the dominant is ovate. The diagonal octagon has a number of axes, or a circle of symmetry (7). Lastly, we have the irregular octagon (8).

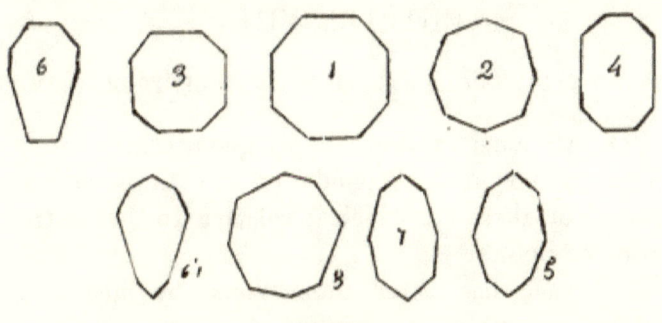

Fig. 33.

89. We could extend this analysis and employ it for polygons of a greater number of sides; it will be sufficient to remark,—

1. That a polygon of an uneven number of sides has but three forms, viz., a *regular and centred*, which has as many axes as there are unities in the number of sides; a *semi-regular* or *even form*, which has an axis of sympathy, and whose dominant is an oval; and the *irregular* or undetermined form (Fig. 34). Fig. 34 shows, successively, the semi-regular forms of the polygon of 3, 5, 7, 9, 11 sides.

2. That the multiples of uneven numbers give rise to resolved forms, and out of these forms rise a submultiple of axes, some even, some irregular.

3. That every even or multiple number gives rise to a number of figures,

Fig. 34.

which correspond to the numerical composition of
the initial number.

CHAPTER VIII.

POLYGONAL FIGURES AND LINES DERIVED FROM THEM.

90. All consideration, simply geometrical, which
refers to polygons, depends on an examination,
general or abstract, which is relative to the distribution of points.

As space has three dimensions, by means of
which a point is determined (seeing that it is
necessary to have two to determine a point on a
surface, and that one is sufficient if the point
is taken as a determined line), the points will be
imagined—

1. Placed one after the other, following a
straight or other line, susceptible of indefinite prolongation; following a circular or other curved line
and closed; and, lastly, following a spiral or other
curved line, with double curvity.

2. Placed the one in proper relation to the
other, on a plane surface or a curved surface
susceptible of being prolonged, independently on
a limited plane surface, or on a curved surface,
limited or unlimited.

3. Separated one in proportion to the other, in
space, constituting a number of figurations, whether
geometrical or artificial.

91. Two points determine an orientation, and
the real or imaginary straight line which binds them

is the rudimentor, being the simplest of all figuration. These points determine a direction.

Four lines in a quadrilateral, bound by the outward distances, form a lozenge, or trapezium, &c., and the distances which bind the opposite points, or outward distances, are diagonals. Five points are placed in a straight line, determining a real or imaginary figure, like a T-form trapezium, cross, or polygon, &c. A number of points distributed in a straight line determine a real or imaginary figure. If the form is real, we get,

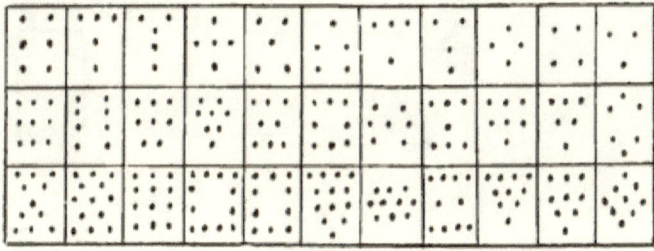

Fig. 35.

successively all the varieties of polygonal figures. If the figure is imaginary, the different arrangements of points are specified by number and situation. It is what appears in a coat of arms.

In the repetition of the figures it is necessary to consider number and situation. Upon situation depends the particular arrangement (Fig. 35), because numbers can be arranged diversely. Generally, the arrangements of points are designed and specified: 1st, by a series which indicate number and situation; 2nd, by the usual figures, determined as the cross and the chevron; 3rd, by the geometrical figures, such as the straight line, triangle, hexagon &c.

92. A determined number of disseminated points give immediately a proportional or corre-

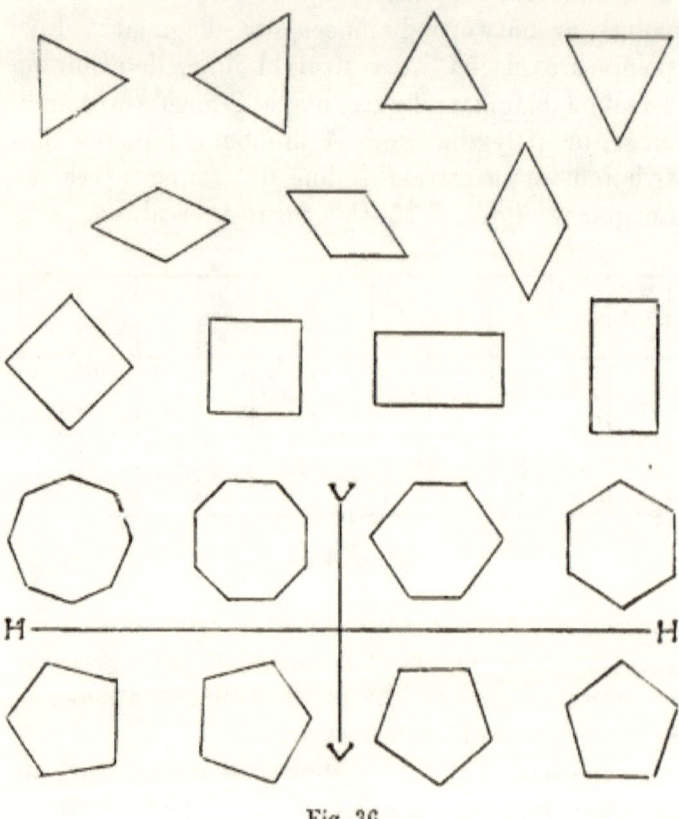

Fig. 36.

lative number of distances. When the number of points increase, the number of distances which bind them increase most rapidly—so three points give three distances; four points six straight lines; five points ten straight lines, and so on. All these straight lines are sides and diagonals of polygonal figures. A distribution of points is symmetrical when the distances are equal, i.e., symmetrically placed; it is regular when the distances are equal, as cut by points equally distant one from the other. This regularity is given in the circular arrangements of the points, from which polygons are or can be formed.

93. Three points, regularly separated, determine the trigon; four points, the tetragon; five, the pentagon; six, the hexagon; and so on, following the ordinary series of numbers.

The trigon can occupy four distinct positions (Fig. 36) :—

1. Based on one side.
2. Stood on the point.
3. Turned to the right.
4. Turned to the left.

It can also occupy several inclined positions, in all of which the triangle is sensibly deformed, the base or point prevailing according to position, and maintaining the integrity of the line. The lozenge gives rise to three distinct figures : stood on the point; cross or lengthwise; and stayed on the side, when the figure is undecided and forms a parallelogram. The rectangle has two positions : the square stood on the side, or on the point; this latter represents a lozenge. The hexagon has two positions as well as the octagon: the first represents the parallelism of the lines and the rectangular directions; the second is only a centred figure, when it is necessary to count the number of sides for determining the polygon. The pentagon can be easily deformed in each of the four cases indicated (Fig. 36), but it is necessary to notice that the changing of position alters only the regularity of form, without giving it a new aspect. The pentagon and other polygons are figures of uneven number of sides, essentially centred.

94. The heptagon, nonagon, decagon, and polygons of a greater number of sides, are centred figures, the natural sense of which is the regularity of the distribution of the sides in proportion to the centre, or following circular order.

CHAPTER IX.

DERIVED POLYGONS.

95. The trigon has no derived figure, nor has the square more than two crossed diagonals. The eight principal polygons, viz., the pentagon, the hexagon, the heptagon, the octagon, the nonagon, the decagon, dodecagon and pentedecagon, have some starred derived polygons of different kinds, as :

The pentagon, a continued starred figure (Fig. 38).

The hexagon, a starred figure formed of two trigons regularly crossed or intersected (Fig. 37).

The heptagon, two continued starred polygons (Fig. 40).

The octagon, a star formed of two squares crossed, and of a continued star (Fig. 39).

The nonagon, a star of three intersected trigons, two continued starred polygons (Fig. 42).

The decagon, a star of two intersected pentagons

—a continued starred polygon—a star formed of two continued starred polygons intersected (Fig. 41).

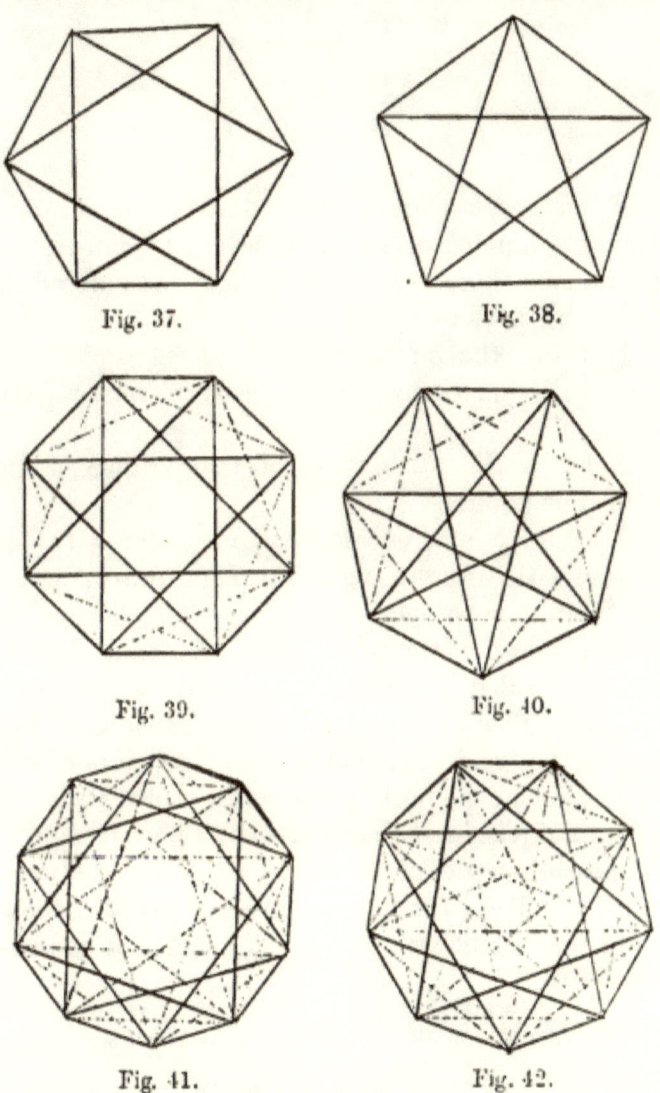

Fig. 37.

Fig. 38.

Fig. 39.

Fig. 40.

Fig. 41.

Fig. 42.

Therefore we get the following table :—

Polygons.	Crossed Polygons.	Continued Polygons.	Starred and intersected Polygons.
Hexagon	1		"
Pentagon		1	
Octagon	1	1	
Heptagon		2	
Decagon	1	1	'
Nonagon	1	1	

From each figure are derived some stars having a number of submultiples and semi-regular polygons ; thus, from the dodecagon are derived the following stars :—

1. From the intersection of three squares, a star with six obtuse points and a semi-regular polygon.

2. From the intersection of four trigons, a star with four points.

3. From a starred· dodecagon, a star with six points, with straight angles turning inward, and a star with four points.

In general, if we take two points on a circumference divided into equal parts, and we join them by a straight line, and then draw a perpendicular line through the middle of that line, all the points of this perpendicular on both sides being joined with the points of division of the circumference, we design all the possible varieties of polygonal forms, between which are placed the figures mentioned before, these being connected with geometrical modes of generation.

INSCRIBED OR CONCENTRICAL POLYGONS.

96. All polygons can be mutually inscribed, whether they correspond by their lines of·diagram,

or whether they are derived one from the other by
the regular junction of points taken on the outline.

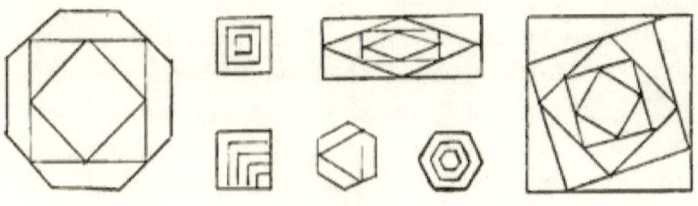

Fig. 43.

CHAPTER X.

DERIVED POLYGONAL PLANE SURFACES.

97. Parallelograms, lozenges, rectangles and trigons are decomposed into small elements of equal

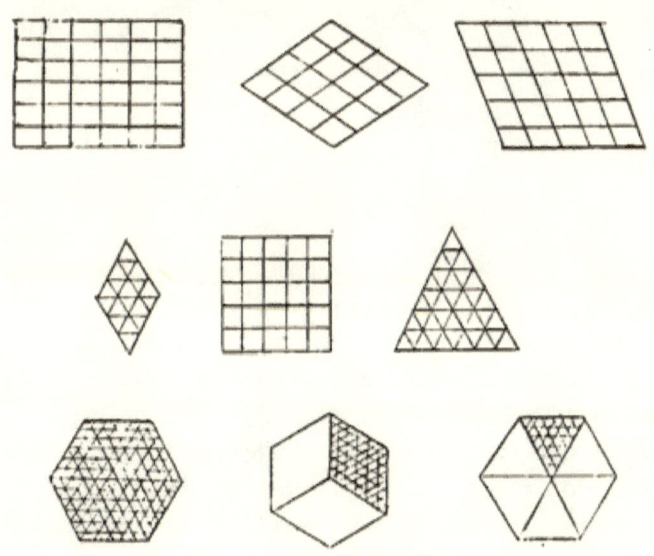

Fig. 44.

forms. This decomposition is established by
the sub-division of the sides in a like proportion.

If the lozenge is trigon, it is decomposed like a
trigon, into elements of the same form; the square
rectangle and the square are decomposed into elements of the same form. This subdivision shows
that the figure is a combination of harmonical network. From the hexagon we obtain a trigonal
network which is or can be decomposed into six

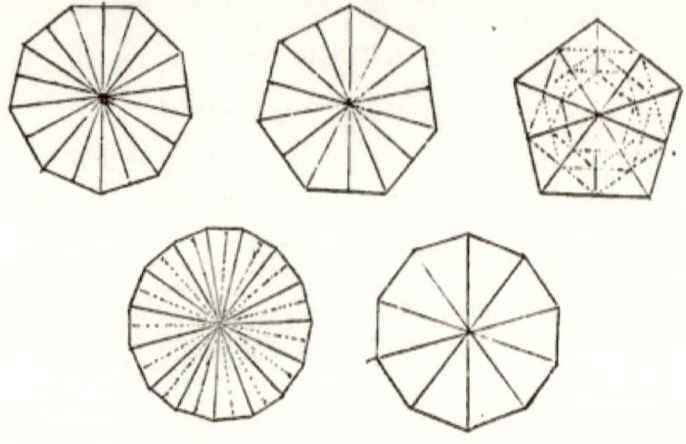

Fig. 45.

trigons and three trigonal lozenges. Polygons,
trigons, squares, the pentagon, hexagon, &c., have
each their own proper net-work, constructed on a
uniform type, the centred polygonal net-work. This
net-work is determined by polygons inscribed parallelly one in the other, with uniform or varied intervals, and intersected by lines coming from the centre.
These net-works may be considered as projections
in the plan of pyramids, whose faces are cancelled
conformably with the net-works of the infinite plan.

Among the modes, infinitely varied by decomposition of the polygonal plane surfaces, it may be necessary to distinguish two fundamental ones. The first (Fig. 45), very uniform and of general application, consists in drawing the rays from the centre to the summits of the polygon, by doing which we decompose the polygon into isosceles triangles, all equal and agreeing with the kind of polygon. The regular or irregular polygonal figures are also decomposed into triangles, varied in manner or form, but not isosceles. The irregular and mid-regular quadrilaterals are decomposed by the diagonal of the figure which binds the opposite summits.

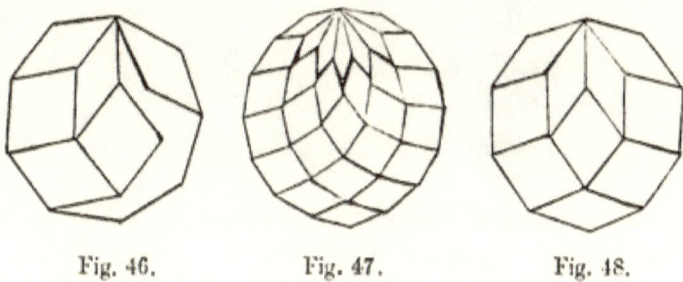

Fig. 46.   Fig. 47.   Fig. 48.

CHAPTER XI.

RHOMBICAL ELEMENTS AND CUNEIFORMS OF POLYGONS.

98. Rhombical elements and decompositions of polygons in lozenges. We may remark that the polygons of an uneven number of sides have a limited number of lozenges, which have an irregular emptiness. On the other hand, polygons of even number of sides are covered entirely by lozenges, placed side by side, and do not leave a void. Moreover, the total number of lozenges is limited to a number of different sizes (Figs. 46, 47, 48).

If the angles of the lozenges are equal fractions of four straight angles, we can arrange them round the same point, and we then get regular stars, whose prominent points have the same size as the laps placed side by side. All the lozenges so obtained have their sides equal to the sides of the generator polygons. It is evident that the triangle and the square have not any derived lozenge. The hexagon has one, and it is the trigonal lozenge. The octagon has two; the pentagon has only one; the heptagon has two, i ts own lozenge and a square. The dodecagon has a proper lozenge, a trigonal lozenge and a square.

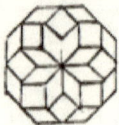   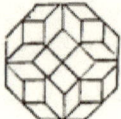

Fig. 49.   Fig. 50.

By repeating circularly the polygons quite decomposed in lozenges, we should obtain some roses similar to those obtained from the octagon (Figs. 49 and 50).

By decomposing the lozenges into smaller ones, we should separate the figure into smaller and smaller parts, which, assembled in every possible way, would determine a great variety of combinations.

99. Two cuneiform elements and decomposition of the polygon in corners. If we join the middle points of the sides of the generator polygon, and the summits of the turning inward angles of the derived polygons to the centre, we decompose the figure into corners, which have for an angle that in the centre of the polygon, and for opposite angles, successively, all the angles of the generator polygon and some starred polygons derived from them.

We have, then, for each polygon a finished

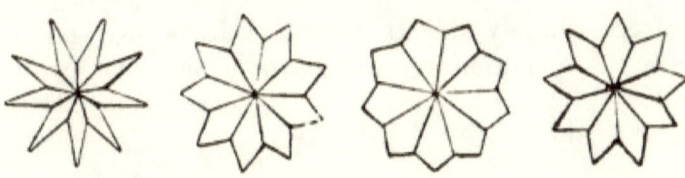

 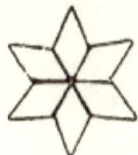

Fig. 51.

series of angles. If we conjugate two and two the angles between them or with the angle of the centre, we obtain lozenges and corners (Fig. 51).

CHAPTER XII.

ASSEMBLED POLYGONS.

100. From the preceding examination we conclude that, on the reverse, some polygonal figures are assembled, and hence determine some composed figures. Two triangles or four triangles will deter-

mine a quadrilateral. From isosceles triangles can
be described convex polygons, and from corners or
lozenges can be described starred polygons. But
for synthetic composition, a condition is absolutely
necessary : it is that the figures have angles in such
a way that, assembled round a particular point,
their sum may be equal to four right angles, i.e.,
placed side by side they do not leave any space.
On this condition we arrange round the same point

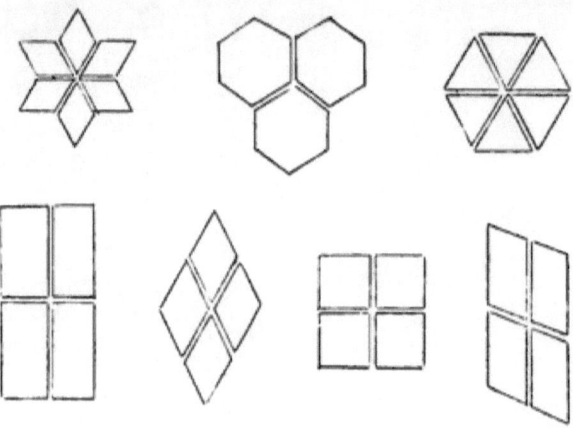

Fig. 52.

a certain number of equal figures, when the angle
of the figure is an exact fraction of four straight
angles, or if those angles together form a sum equal
to four right angles.

We can place round a point six trigons, six or
three trigonal lozenges, three hexagons. The parallelograms, the rectangles and the squares are
arranged four together, round a point (Figs. 52
and 44).

For the different figures the fundamental assemblages are as follows:—

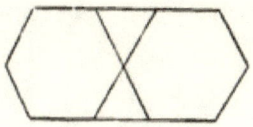

Fig. 53.

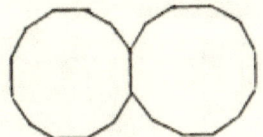

Fig. 54.

1st. 2 hexagons and 2 trigons (Fig. 53).
2nd. 2 dodecagons and 1 trigon (Fig 54).

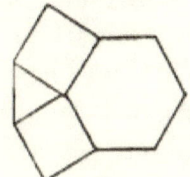

Fig. 55.

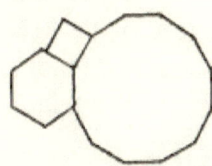

Fig. 56.

3rd. 1 hexagon, 2 squares and 1 trigon (Fig. 55).
4th. 1 dodecagon, 1 hexagon and 1 square (Fig 56).

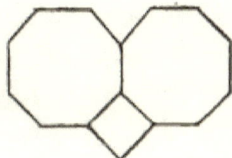

Fig. 57.

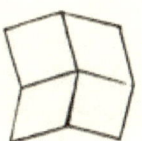

Fig. 58.

5th. 2 octagons and 1 square (Fig. 57).
6th. 2 lozenges and 2 squares (Fig. 58).
7th. 2 rectangles and 2 squares (Fig. 59).
8th. 2 lozenges and 2 rectangles (Fig. 60).
9th. 1 decagon and 2 pentagons (Fig. 61).

10th. 1 pentedecagon, and 1 decagon, and 1 trigon (Fig. 62).

Fig. 59.   Fig. 60.

A hexagon surrounded by six squares and six trigons gives a dodecagon. The reunion of the dodecagon with the trigon is equal to the assemblage of the four figures of the hexagon, the square, the trigon, and the trigonal lozenge.

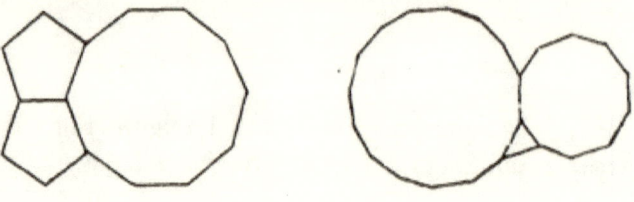

Fig. 61.   Fig. 62.

The sixth assemblage is changeable, according to the accent of the lozenge. The seventh is changeable, according to the proportion of the rectangle. The eighth is changeable, by the simultaneous variation of a lozenge or a rectangle.

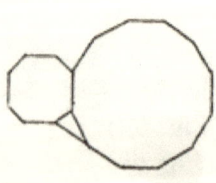

Fig. 63.

These eight assemblages can be repeated indefinitely upon the whole surface of the plan.

The ninth assemblage is finished, and composed of a decagon with a crown of pentagons. The tenth assemblage is equally finished, and composed of only four figures, a pentedecagon, a decagon and two trigons.

We can arrange around the same point a dodecagon, a mid-regular octagon whose axes are transversal, and an isosceles triangle. (See Fig. 63.)

101. The polygons derived by the prolongation

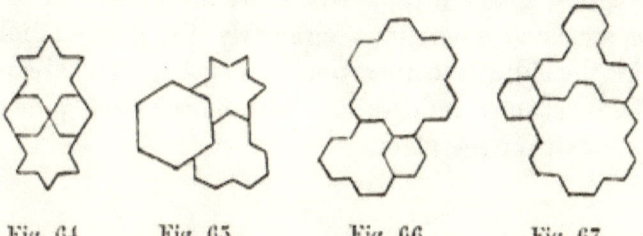

Fig. 64.　　　Fig. 65.　　　Fig. 66.　　　Fig. 67.

of the angles can be assembled, and give rise to combinations infinitely varied. We shall take, for example, some of the figures derived from the hexagon and from the octagon

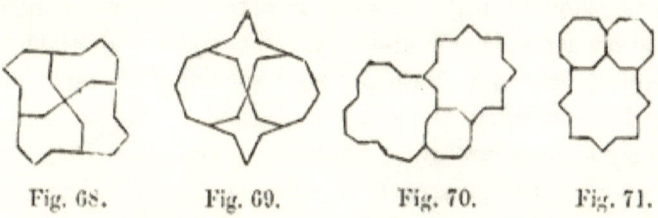

Fig. 68.　　　Fig. 69.　　　Fig. 70.　　　Fig. 71.

For the hexagon we join two hexagons and two hexagonal stars (Fig. 64); 1 hexagon, a star and 1 dodecagon with three inward-turning angles (Fig. 65); 1 hexagon, a hexagonal pearl and a rose with six hexagonal projections (Fig 66); 1 hexagon, a rose with seven assembled hexagons, 1 pearl with four assembled hexagons (Fig. 67).

For the octagon, we join 4 octagons with two

inward-turning angles (Fig. 68); 2 octagons and
2 stars with four points (Fig. 69); 1 octagon, the
starred octagon of two squares, and the polygon with
16 sides with four projections (Fig 70); 1 starred
octagon, 2 octagons, and ultimately a square (Fig. 71).
This assemblage, reduced to the star and to the two
octagons, gives a final disposition, composed of
the star and 8 octagons, circularly distributed; but
with the aid of the insertion of a square, the assemblage becomes indefinite, and applicable all along
the extent of the plan.

CHAPTER XIII.

DERIVATIVE POLYGONAL DECLINATIONS.

102. It is necessary to conceive the outlines of
polygonal figures as formed by two means: the
first, similar in its linear course, treats of lines which
follow each other under any inclination; the second,
resembling the diagram of disposition or circular
order, considers the sides or angular folds as regularly divided around a point or centre.

I. General construction of the polygonal delineations. The irreducible and last elements of the
polygonal delineations are the angles and the sides
or segments of straight lines.

Those segments follow each other by making
angles between them. The segments are changeable in linear sign and the angles are changeable in
angular sign, &c. It is necessary to consider,
separately, the angles and segments which fix the
limits of these angles. All segments can be uniform

or varied, which would give rise to four modes of generation of polygonal declinations.

1. Equal segments and equal angles. Starting from a fixed segment, or from a fixed angle, then tracing or repeating successively the segment and the angle, we obtain, generally, two kinds of polygonal figures; regular convex polygons, or continued starred polygons. If, for example, it be the one in the octagon, we should get the convex octagon; if the angle

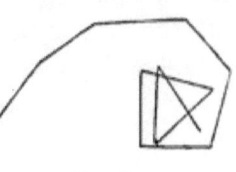

Fig. 72.

is the one in the starred pentagon, we should get the starred pentagon.

Equal Segments and Varied Angles.

2. If the uniform segments follow each other by reclining under angles, successively smaller and smaller or greater and greater, we obtain a declined polygonal outline, which inscribes or circumscribes a bent line (Fig. 72). The angles can be varied in

Fig. 73.

every way, whether uniformly or by a regular declination, or by a varied way, the angles being equal two and two, &c., &c.

3. Equal angles and varied segments. The segments regularly decreasing in size while they

incline always under the angle, determine a
re-circled polygonal outline, which inscribes or
circumscribes a curved or rolling line. If the
segments were equal two and two, the angles would
be straight, and the outline would inscribe a rolling
with an equal spirular and uniform interval.

If the angles are those of the trigon, square,
pentagon, hexagon, &c., we obtain rectilinear
rollings, which are inscribed in these figures
(Fig 73).

Varied Angles and Varied Segments.

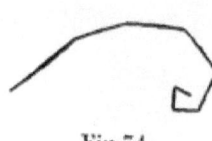

Fig 74.

4. Whilst the angles vary
in size, the segments vary in
size also, and the figure describes a bent line, which goes,
being prolonged as far as the
rolling, according to the degrees of progression (Fig.
74). This declination describes a curl, and the
progression is continued in the opposite way
according to the variation of the angular sizes.

By prolonging the segments in every way, we
should get complicated declinations which correspond to.the starred polygons.

II. General construction of the rectilinear and
curvilinear polygonal delineations. The sides of
the polygons belong by halves to each angle. We
will consider these polygons as determined by the
circular reparation of fold, whose axis of symmetry
agrees with the rays of the diagram of disposition ;
a general outline being traced, we get a number of
figures derived by the abatement of the angles,
which we are going to show in two examples of the

octagon and hexagon (Fig. 75). Starting from the octagon, if we abate two opposite folds, we get (1) ; if we abate the four folds we get (2) ; if, taking the middle points of the sides, we abate the eight angles, we get (3) ; if we only abate $2_{and}^{in} 2$ we get (4). The figure (5) is obtained from (2) by the re-abatement of all the angles turning inwards ; (6) is obtained from (3) by the re-abatement of the angles two in two; and (7) is obtained from (3) by the re-abatement of all the angles. The figure (8) is made from (7) by the re-abatement inside the projecting

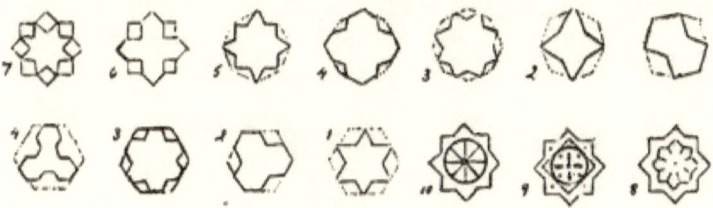

Fig. 75.

angles ; and (9) is obtained from (7) by the re-abatement of the straight angles turning inwards.

Starting from the hexagon, and taking the middle points of the sides, by abating the angles we get the figure (1) ; abating them only two and two, we get (2). Separating the sides in three equal parts, and abating the angles, we get (3), (4), &c. From the dodecagon, and generally every other polygon, we obtain infinitely varied figures ; but the detail would be too tedious, and not absolutely necessary.

103. If after having divided the sides of the figure into two parts, we prolong the outline in a straight line, we get a series of broken lines, which equal a series of angles, placed one after the other. On the contrary, a rectilinear gear, being perpendicularly folded, following the kind of angle, will form a convex polygon or starred polygon. These gears, having an axis of symmetry, being in the straight line, the circular outlines will then have a head circular line, which is a polygon of somewhat angular form. We now perceive the bending of the linear order with the circular order, and all the ranges, lines and gears will have straight and circular headlines.

Fourth Section.

EXTENT AND CORPOREAL FORMS.

104. We get the perception or immediate intuition of the corporeal forms which exist in the extent by fixing and describing them; we conceive them as determined by the subordination of the extent to the co-ordination. These co-ordinate forms are of several kinds : the mathematical co-ordinations, which determine size; the geometrical co-ordinations, which determine the kind or sort, by the consideration of order and place; the æsthetical co-ordination, which determines the æsthetical sense of the form ; the harmonical co-ordinations, which refer to the individuality of the forms.

Homogeneous extent has three dimensions of the same nature which are essentially uniform ; the size of these dimensions measure extent ; to these three dimensions correspond three co-ordinations, which determine the order and place of extent; such are the mathematical co-ordinations, numerical and geometrical.

Extent, in the æsthetical sense, has three dimensions which determine its size; to these three dimensions, which characterize the forms of the extent, correspond three co-ordinations, which determine the order, place, and sense of the extent. These three co-ordinations are essentially uniform, they are height, breadth and depth. Such are the geometrical co-ordinations, as well as mathematical and architectonical, which determine size and form.

The particular forms of extent give out two kinds of co-ordinations, which essentially determine them; the geometrical co-ordinations or generations of form and the simple co-ordinations are harmonical to the æsthetical kinds of form.

105. If it is easy to trace curved lines or figures on a plan containing a known number of points, there is no necessity for drawing that figure which is so well known, and finished, as subject to the laws of generation. This is termed descriptive geometry, the object of which is to change the plane surface by graphically operating upon that plan.

A modelled surface requires in every degree continued varieties of curvity, varying in size and accent. All these variations follow each other, and determine a series of concave or convex undulations. On such a surface we can conceive a number of lines with continued and inflected curvity, more generally plain curved, with double curvity.

106. Amongst arbitrary modelled surfaces we can distinguish some named, known and familiar to us.

The geometrical definition of a surface consists in ascertaining the law of description of the surface, by means of a line, whether the line may simply move in space without changing form, or whether it may change form at the same time that it moves.

Generally, we call headlines those upon which rests the generating line for describing a determined surface.

First Class—Ruled Surfaces.

These plane or curved surfaces are produced by the moving of a straight line. This class contains:—

1. The order of the developing surfaces.
2. The order of the clumsy surfaces.

The order of surfaces which are developing contain:—

1. The series of cylindrical surfaces.
2. The series of conical surfaces.

Second Class—Enveloping Surfaces.

This class principally contains :—

1. The order of changing surfaces.
2. The order of enveloping surfaces.

The ordinary straight cylinder is a ruled surface, because we can conceive it as formed by a straight generating line which moves parallel to itself, following an outline or circular headline ; and also an enveloping surface, because we can conceive it as formed by the uniform progression of a circle or sphere following a straight headline. The cylinder is a developing surface because we can put down the surface on a plan; it is also a changing surface, as it can be described by a straight line or a rectangle, &c.

107. Here is another classification, equally artificial, but less abstract.

All surfaces, and generally all forms, can be conceived as formed by the progression of a line or plane figure following a headline. This progression is uniform or declined or varied, following the different profiles which guide and circumscribe the extent of the generating movement; and again, those forms are referred to a plan of construction or a symmetrical co-ordination.

CHAPTER I.

FOLLOWED OR LINEAR FORMS.

108. All these forms are produced by the uniform progression of a generating line following a headline. The generating can be :—

1. A straight line, an arc, a curved line, a handle, &c.

2. An open or closed polygonal figure.

3. A curved figure, simple or varied, a circle, an arc, an ovate, an oval, &c., an inflected or turned rose, &c.

4. A plane delineation varied in an authorized manner.

The headline can be a straight one, an arc, a circle, an oval, a spiral, a declined spiral.

The conjugation, two and two, of a generating or headline produces a number of forms; for example, a generating straight line and a head straight line produce a plan; a circular generating and a head spiral line produce the serpentine or twisted column, &c., &c.

CHAPTER II.

DECLINED FORMS.

109. These forms are produced by the declined progression of a generating line following a headline.

The generating and head lines are equal to the preceding, only the principle of declination intervenes for the determination of the forms. So, taking a straight headline and a circular generating line, and guiding the progression by a straight inclination of the headline, we obtain a cone, whose volume declines from the opening of the angle to the summit. The progression taking place along a curve or arc, every other condition being the same, we get an ovoid form. If starting from a circular generating and from a spiral headline, and the progression be declined instead of being uniform, we get instead of a cylindrical serpentine a conical serpentine.

SYMMETRY IN THESE TWO CLASSES OF FORMS.

110. Symmetry in solid forms refers to a point or a straight line, or one, two or three plans which are rectangularly cut. By the point, a number of plans pass arranged in every possible way. By the straight line pass an equal number of plans, but all these plans containing the straight line are directed in the same way. Thus the sphere and the regular polypedrons have centres of symmetry; the cylinder, the cone, the prisms, the pyramids, &c., have centres of symmetry. The ovoids and the ovaloids have one, two or three plans of symmetry.

From that and that which concerns the cylindri-

cal and conical forms, we notice generally that if
the headline is straight and the section a generating
circular, it passes by the straight line in a number
of plans; if the headline is straight, and the generating polygonal or varied, it passes by the straight
in a number of plans. If the headline is an axis or
curve, there is only one plan of symmetry which
passes by the headline. If the headline is a spiral,
there is not any symmetry, but only an imaginary
axis round which is rolled the solid.

CHAPTER III.

SPHEROIDAL FORMS.

111. The sphere is an absolute form whose surface
and solidity are strictly referred to a point; it is
then an unique form whose every examplar is
identical. This does not help us to conceive it as a
surface of revolution produced by the rotation of a
circle, or as a conical surface described by the progression of a circle following the rectilinear headline
or a circular declining line.

But if the generating of circular, ovate or oval
form is moving, following a finished line, the declining line may be a circle, an ovate, or an oval, and we
obtain the numerous variety of the ovoid having one,
two or three plans of symmetry, or a number of plans.

CHAPTER IV.

VARIED FORMS.

112. All cylindrical, conical and ovoid forms are
produced by uniform and declined progression, fol-

lowing arcs or curved lines; but if, instead of these simple lines, we employ varied declining lines, which should contain a number of fundamental affections of the lines, we should obtain a number of varied forms, in general, such as the diapered forms, the essential parts of the vases parts of architecture, &c., &c.

According to the relative situation of the head and generating lines, and the reciprocal nature of their form, we obtain the elements of certain forms essentially different. For example, taking for the head a straight line, for generating a circle, and for declining or profile a curved line, according as that curved line turns its concavity or convexity on the side of the headline, we obtain a conical, convex, or open form. In the torns we get these two systems of curvity; in the ordinary cylinder all the sections, perpendicular or slanting to the headline, are curved lines. All the sections, on the contrary, which pass by the headline, or are parallel to it, are straight lines.

From that we distinguish three kinds of curved surfaces :—

1. Those which have all their curvities in the same way, starting from the same point, as the sphere, the oval, and generally the convex surfaces.

2. The curved surfaces which have their curvities always in the contrary way, as the torns, the serpentine, and generally the inflected surfaces.

3. Those surfaces simply rounded, or having curvity only in one way, as the cylindrical or conical surfaces.

I

113. The superficial extent has two dimensions and two co-ordinations; the corporeal extent has three dimensions and three co-ordinations. A dimension predominates, or is harmonically lost in giving the circle; again a dimension predominates, the two others being lost or distinct, or the three dimensions are lost harmonically in the sphere. The fundamental forms of the linear extent are the plan, circle, label and angle; the fundamental forms of corporeal extent are the cube, sphere, and the prismatical parallelogram. The segments of the fundamental forms of extent which are superficial are the triangle, the quadrilaterals, and polygons. The dominant of the plane surfaces are circular, quadrangular, linear and angular; the dominant of extent are cubical, spherical, prismatical or linear, polyhedrical or pyramidal. In general the regular polygons correspond to the regular prisms, the regular pyramids, and the spherical polyhedrons. The starred polygons correspond to the prisms and pyramids with sides, and to the starred polyhedrons which are produced from the ordinary polyhedrons by diagonal plans, as well as to the starred polygons which spring from the ordinary polygons by diagonal lines. The circle corresponds with the cylinder and the cone or the sphere. The outlines of the plane surfaces correspond with the surfaces of the solids.

114. All the preceding theoretical considerations belong to elementary and general geometry.

115. In the arts and handicrafts the surfaces and forms are diapered directly, and plainly, by modelling or sculpture. The mode of manutention is more or less definite, according as the form is more or less definite; but all the mechanical improvements given to the tools, or engines, do not add anything to the notion of the form. It is not in the industrial improvement of the mode of elaboration that the development of the art of form lies; on the contrary, this improvement, which is quite in the economical sense, and organized for a rapid, abundant production, is really prejudicial, though it assists the power of invention and creation.

REGULARITY AND SYMMETRY.

116. A drawing, whether it be touch, delineation, figure-form or ornament, is simple, composed, complex or composite.

A drawing is simple when we can separate it, still keeping a form, or when it is the starting points of compositions or arrangements, or when this form is always found in definite drawing. It is composed when we can distinguish distinct parts which assist in its formation, but so united that we cannot separate them without breaking the unity of form. It is complex when it is formed of different parts, and the bond which unites these different parts is apparent. It is composite when the parts determining it are of different sorts, and the part is prominent or dominant.

117. A form is regular or irregular. It is regular when it is constructed in accordance with an apparent law, when we can distinguish regularity in the arrangement of the disjoined parts, and when we can conceive a harmonical bond in the arrangement of its different parts artificially separated or only distinguished by analysis.

The idea of regularity has two distinct acceptations, according to what they depend on. The harmonical bonds which unite the different parts of the disposition, and which consist in a regular declination, are :—

1. The position following the curved line.
2. The size of the parts.
3. The accentuation of form, which is of a special nature, and does not depend upon the idea of a developed law, for which could be substituted sensible and immediate form.

A regular form can be symmetrical, declined, or varied, so that the ornamental regularity is separately or conjointly determined—

1. By the symmetry.
2. By the declination.
3. By the variation.

CHAPTER 1.

SYMMETRY.

118. A form, whatever it may be, has its constituent parts referred to a diagram, also its geometrical plan of construction, or this form is repeated in proportion to a geometrical diagram of disposition. Accordingly as the parts of a figure or disposition are assembled at a point, line or plan, the symmetry differs.

The solid forms of uniform constitution have all their parts drawn up in a point, as in the polyhedrons and spheres ; in the straight line or axis, as the prisms, pyramids, cylinders and cones ; in one or several plans, as the sphere, cube, ovoids, &c. Let us now examine the different cases of symmetry without separating them from the forms which contain them.

119. Seeing the forms in their formal unity and in the integrity of their image, we are obliged to distinguish them, and to define them, in the following manner: a form is uneven, even, diagonal or quartered, revolved or radiated, ternary or senary.

1. Uneven forms. A form is uneven when it has neither axis nor centre of symmetry. An uneven form is regular or irregular; the regularity consisting in a progressive and harmonical variation

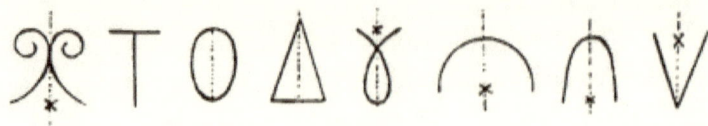

Fig. 76.

of what constitutes the essence of form : for example, a curved line is regular because there is a continued change and progressive modification in the curvity. An uneven form is simple or compound, being simple when its principle of unity holds entirely to its essential nature ; thus a curved line is a simple form, because there is continuity in the declination of the curvity. An even Greek parallel is complex, because the parts which compose it are different and form each a unity or simple form. To speak correctly, it is not a form but a disposition.

2. Even forms. A form is even when it has an axis of symmetry which separates it into two equal

parts, and these correspond by turning. The arcs, handles, ovals, straight angulations, isosceles triangles, &c., are even forms, and the simplest of all. Generally, symmetry separates the object into two parts, places in the middle the unique parts, besides those which are repeated, and which form a sort of equilibrium, giving order, liberty and beauty to the object. (Fig. 76.)

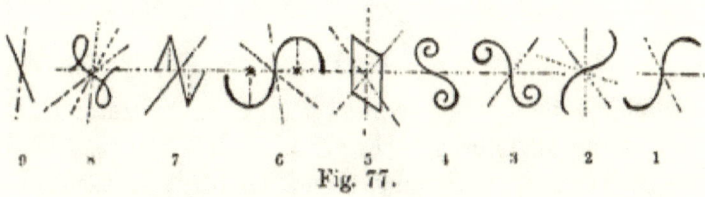

Fig. 77.

3. Diagonal forms. A diagonal form is constructed by reference to a point : it has its halves placed in the inverted way in proportion to one of the straight lines, which pass by the centre or symmetrical point. In general, a diagonal form is

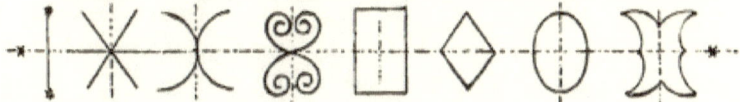

Fig. 78.

determined by the doubling of an uneven or even form ; it has always a figure which repeats itself by invertion, Fig. 77 (6), (7), (8), (9).

4. Quartered form. The two even and diagonal symmetries together determine the quartered symmetry. A quartered form has two symmetrical axes which are crossing at the point of the symmetrical centre. (Fig. 78.)

There is a straight symmetry in cross and diagonal figures (symmetry at cross-wise). The straight, the cross-wise, the rectangle, lozenge, oval, &c., are forms as well quartered as are those of diagonal compositions.

5. Quarternary form. A form is quarternary

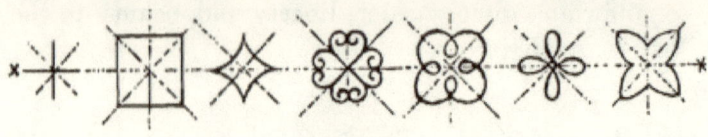

Fig. 79.

when it has four symmetrical axes crossing or radiating at a point. (Fig. 79.)

6. Ternary or Senary form. A form is ternary or senary when it has three symmetrical axes which cross at a point or which radiate from this point in number three or six (Fig. 80). A senary form has for form-envelope a hexagon or a circle. A ternary form has for form-envelope a trigon or a circle. The quarternary, ternary, senary, quinary

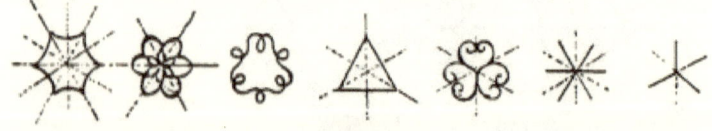

Fig. 80.

forms, &c., are forms whose pieces are circularly turned down; so also is the quartered form, but this last implicates two axes rather than a centre and rays, in the case where the number of pieces is uneven, or the multiple of an uneven, and a centre and axes in the case where the number is even, or the multiple of an even number.

7. Revolved form. A form is revolved when it

has all its parts circularly arránged round a real or imaginary point, and turned in the same way (Fig. 81). A revolved form is an uneven one if the number of pieces are uneven; the imaginary point is then a centre of disposition, seeing that the figure has neither axis nor symmetrical centre. A revolved form is of diagonal symmetry if the number of pieces is equal or even, the centre of disposition being then a symmetrical centre by which passes a number of axes. Generally, a diagonal form is one composed of two pieces.

8. Radiated form. A form is radiated when the symmetrical axis is an indefinite number too

Fig. 81.

great for counting. The simplest figure is the circle which has a number of axes.

120. Every even, uneven, diagonal, or quartered form has a degree, the idea of the figure-envelope which contains it, and which can be an irregular figure, or more particularly the isosceles triangle, the quadrilateral, triangle, or the circle. Every revolved and radiated form (i.e., every form circularly constructed round a point) implicates a figure-envelope the image of a regular polygon, from the trigon to the circle. For seeing the question in all its generality it is necessary to consider two very important points :—

1. The different positions which a determined form can occupy.

2. The reciprocal positions or conjugations of

two identical or different positions of a determined form.

I. Positions of the Forms.

121. 1. Radiated forms. A radiated form, having more than four axes of symmetry, and whose form-envelope is a polygon of more than eight sides, or a circle, is not susceptible of direction, and remains indifferent.

2. Revolved forms. A form revolved of one piece is uneven ; a form revolved of two pieces is uneven and diagonal; a form revolved of three pieces, having for form-envelope a triangular outline, can occupy eight positions, as the triangle. A revolved form of four pieces, having for form-envelope a quadrilateral, can occupy four positions.

3. Quarternary and ternary forms. A quarternary form, which has four axes of symmetry, and for form-envelope an octagonal or circular outline, can occupy two positions. A quarternary form or senary form, which has three symmetrical axes, can occupy eight positions, as : an axis being horizontal, the points turned on the right or on the left; an axis being vertical, the point turning up or down; and an axis being oblique, in two positions, on the right or on the left, can get a point turned in one way or in the other.

4. Quartered forms. A quartered form, or one with two axes, has two positions if the form-envelope is a square ; or four positions, one horizontal, the other vertical, and the other two oblique, if the form-envelope is a lozenge or a rectangle.

5. Diagonal forms. A diagonal form, having a

symmetrical centre or a number of axes, occupies the positions which depend on the form-envelope. The form being generally rectangular or a lozenge, occupies four positions, one horizontal, one vertical, the other two oblique.

6. Even forms. An even form, having only one axis of symmetry, i.e., two equal faces on each side of the axis, and two different faces in the way of the axis, can occupy eight positions, as: the axis being horizontal, the form is turned; or on the other, the axis being vertical, the form is turned up or down; the axis being oblique on the left, the form is turned in one way or in the other ; and the axis being oblique on the right, the form is turned in one way or the other.

7. Uneven forms. An uneven form has four different faces, while the even form has two only. It occupies, then, a double number of positions, i.e., sixteen, that is, if the form-envelope is lengthened or triangular. This form-envelope being almost circular, as in the rolling or quadrangular, or the rectilinear rolling of the Greek, the form would occupy only eight principal positions.

II. Conjugations of the Forms.

122. By combining or putting together two and two, the positions or directions of a form, we obtain the conjugations. It is necessary to consider in the conjugations :—

1. The reciprocal directions. One form being directed in one way, while the other is directed in the same or in a different way.

2. The reciprocal positions. The two forms being considered only for their directions, the conjugation two and two of these directions determines the reciprocal positions.

3. The reciprocal situations. A form occupying a determined place, the second form, which is conjugated, occupies a different place, which is, with the first one, in a relative situation whatever it may be.

It is the reciprocal position which determines the sort of conjugation, the number of directions or the kind of form determining the varieties of the conjugation pertaining to this form; and the reciprocal situation determines the figure or actual form of the conjugations; and more,~they obtain varied images, according as the two forms may remain different, or separated by an interval, whatever it may be, or are touching one another.

1st. *Uneven forms.* Taking for example the rolling, and putting together the different positions of this form, we obtain from them six sorts of definite conjugations :—

1. The uneven conjugation, where the two pieces have not any definable relation.

2. The revolved conjugation, uneven or unsymmetrical, but regular, because the points are circularly corresponding together by proportion to an axis or curvilinear diameter.

3. The followed conjugation. Two forms being repeated, the distances of the corresponding points are all equal and parallel between them, i.e., these distances are directed in the same way and following a linear head-line (Fig. 82). This conjugation is equal to its precedent; in one case the forms are

circularly followed, in the other, they are linearly followed.

4. The diagonal conjugation. Two forms being diagonally repeated, or in situation reciprocally contrary, have all the distances of the corresponding points equal and are crossed in the middle by the same point which is the symmetrical centre (Fig. 84).

5. The conjugation in the wrong way. Two forms being symmetrically repeated by proportion to an axis have all their corresponding points equally distant from this axis. The distances are all parallel and crossed in their middle by the symmetrical axis which is perpendicular to them (Fig. 83).

6. The contrary conjugation. Two forms being

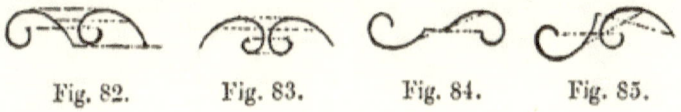

Fig. 82.   Fig. 83.   Fig. 84.   Fig. 85.

repeated and contradicted, have their corresponding points equally distant, but those distances are not all directed in the same way; they are sometimes parallel, sometimes crossed in a changeable manner and in different places (Fig. 85). It is necessary to observe :—

1. That a followed conjugation is uneven, i.e., has neither axis nor symmetrical centre.

2. That a conjugation in the wrong way is even, i.e., has a symmetrical axis.

3. That a diagonal conjugation is diagonal, i.e., it has a symmetrical centre.

4. That a contrary conjugation is uneven, but has a transversal axis or a diameter which separates the figures in two equal parts as quantity, and superposable by moving and turning.   Each of the conjugations is, for situation :—

1. Separated (Fig. 86).
2. Contiguous (Fig. 87).
3. Crossed (Fig. 88).

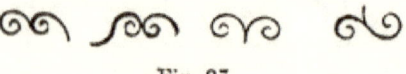

Fig. 86.

They can be defined simply and briefly by saying of a form that it is followed, even, diagonal, or contrary; and more, add that each of them is separated, contiguous, or crossed.

Fig. 87.

To construct these different conjugations, it is necessary to place the initial form in proportion with the square touch.

Fig. 88.

2nd. *Even forms.* An even form has its axis vertical or horizontal; if the position of the axis is oblique, the form can be considered as uneven. The symmetrical axis can occupy some positions and directions which are changeable: we obtain, by

the reciprocal combination of those positions and
directions and following the relative situation of
the parts, the following conjugations :—

123. I. *Axes are parallel, straight or oblique.*

1. Directed in the same way (Fig. 89).
2. Directed in the opposite way (Fig. 90).

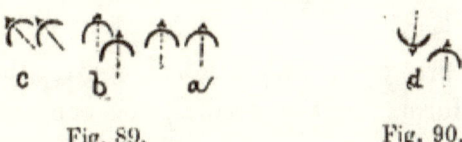

Fig. 89.   Fig. 90.

The first follow in the wrong way if the forms
are directly corresponding, *a*; the second follow
and alternate if the forms are obliquely following
one another, *b, c*. The first is an even form, the
second an uneven form. The third conjugation, *d*,
where the axes are turned in the contrary way,
is always diagonal, whatever the relative situation
may be.

Fig. 91.   Fig. 92.

II. *Axes are in prolongation.*

1. Directed in the same way (Fig. 91).
2. Directed in the opposite way (Fig. 92).

In the first case the conjugation is followed, and is
an even form; in the second, it is the backward
conjugation, and a quartered form.

III. *Axes are perpendicular or oblique one with the other.*

1. Directed in the same way (Fig. 93).
2. Directed in the opposite way (Fig. 94).

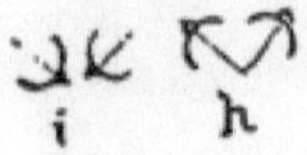

Fig. 93.    Fig. 94.

In the first case, the backward conjugation, it is an even form; in the second, the conjugation is revolved. And in the opposite way, if starting from

Fig. 95.

a form determined in its position, it is conjugated contrarily; revolved in every other case.

3rd. *Diagonal forms.* A diagonal form determines only two conjugations, the followed and the backward conjugation. The other uneven conjugations,

Fig. 96.

b or *a* (Fig. 95), being crossed by the symmetrical centre, determines a revolved or quartered form.

4th. *Quartered forms* (Fig. 96). Two quartered forms, having a big and small axis, can have the axes parallel, and the conjugation is then followed. If the axes of the same are perpendicular the one with the other, the conjugation is then in the opposite way ; and if the forms are crossing by the centre of the figure, we get a quarternary form. When the axes are oblique the one with the other, we get the symmetrical or opposite conjugation.

5th. *Revolved forms.* Revolved forms are uneven or diagonal. Uneven revolved forms of more than five pieces are almost indifferent, and determine the followed and opposite conjugations. The revolved form of three pieces determines all the conjugations of the uneven form; the simplest, the opposite conjugation, being regularly crossed, determines a senary form.

6th. *Ternary and Senary forms.* One of these forms which has less than four symmetrical axes, gives rise to the followed and reciprocal conjugations. A ternary form, or one with three axes, which has a triangular form-envelope, must be considered as even. A senary form which has a hexagonal or circular form-envelope, gives rise to two followed and opposite conjugations.

7th. *Radiated forms.* A radiated form can only be submitted to the followed conjugation, or to the principle of repetition ; it is and remains indifferent in every situation, being exclusively constructed by reference to a centre. In general, and considering all the forms as having at least three axes, a radiated form of 3, 5, 7, 9 . . . . pieces is even; a radiated form of 6, 10, 24 . . . pieces is quartered; a radiated form of 4, 12 . . . pieces is quarternary ; a radiated form of 8, 17. . . . pieces is radiated. ‘

K

Part III.

ORDER AND DISPOSITION IN ORNAMENT AND FORMS.

124. Generally, the differences in the modes of construction of the ornaments have importance only by their relation to the proper genius of mankind, and their neatness and logical or geometrical strictness are not the measure of superiority of one style or another. This appears from the following brief comparison of three particular styles of ornamentation.

1. *Japanese* ornamentation is disjoined without regular co-ordination. It is composed of forms taken from Nature, arts, and society. All these forms, which have very accentuated profiles, are mixed in confusion. It is the habit of these people to give the forms of ornamentation neatness and precision, which are the modes of perception and consisting in the immediate determination of forms in plane surfaces and under a correlative graduation.

2. The ornamentation of the *Asiatic* is regular and continued upon all surfaces. It is a multiplied and luxuriant ornamentation, which is not strictly disorderly, but which yet is not quite orderly.

3. The ornamentation of the *Greek*, or more generally the classical ornamentation, is essentially ordered and regular. This ornamentation is contained and distributed under the form of ranges, which follow the great lines of edifices.

CLASSIFICATION OF DISPOSITIONS.

125. Forms and ornamentation regularly constructed have all their varieties divided in two classes.

FIRST CLASS : UNLIMITED DISPOSITIONS.

1. Order. Dispositions in series.
2. „   Multiplied dispositions.
3. „   Agglomerated dispositions.

SECOND CLASS : LIMITED DISPOSITIONS.

1. Order. Figurative dispositions.
2. „   Co-ordinate dispositions.
3. „   Constructed dispositions.

UNLIMITED DISPOSITIONS.

126. This class contains three orders of disposition, as :—

1. The dispositions in series or in proportion to a headline; the forms or integral elements being placed together following a line of indefinite nature, as the straight, the spiral, or the circumference, but with uniform course.

2. The multiplied dispositions, proportioned to the co-ordinated plan, from which they take the two dimensions; the forms have proportion between them, or multiplied lines.

3. The agglomerated dispositions, or co-ordinated solids, which take the three dimensions of space.

CHAPTER I.

ORDER OF DISPOSITION IN SERIES.

127. The dispositions in series are proportioned to a rectilinear headline, curvilinear and voluble; they consist in the indefinite succession of identical or different forms. The dispositions in series or lines are obtained by the repetition of a number of forms, these forms being identical with or different from one another. These different rhythms, taken separately or mixed together, determine a number of dispositions in seri es.

Fig. 97.

1. *The repetition.* The repetition is the parallel or followed succession of a form in line, or of a

Fig. 98.

linear affection in the course of a delineation (Fig. 97).

2. *The alternation.* Two different forms which

Fig. 99.

follow one another determine an alternation (Fig. 98). The alternation is simple when the forms follow one and one; it is compound when it exists by equal groups in number and extent; it is varied when the groups are not equal, and this inequality can go as far as the intercalation.

3. *Intercalation.* We shall name as intercalation every disposition, series, line coming from the

Fig. 100.

repetition, intersected, which is a different form. This form could be repeated in its turn, and would produce alternations of groups equal and unequal (Fig. 99).

Fig. 101.

4. *Period.* This is the repetition of groups, composed at least of two, three or four different forms. If the composing group has only two forms, the period is an alternation. The regularity of these dispositions consists in the parallel and followed repetition of the composing group (Fig. 100).

Fig. 102.

5. *Recurration.* The periodical disposition is not · symmetrical; it is however a regular disposition, because there is a uniform repetition of a group of forms following in the same order. The recurra-

tion has a period, or a changeable cut, with a number of forms (Fig. 101).

Fig. 103.

Four forms give a period of eight elements; one form being repeated four times, another twice, the other two occurring only once. This cut is com-

Fig. 104.

posed of two periods, which is equal to an alternation of the two groups (Fig. 102).

The disposition can be infinitely varied. Thus the recurrence Fig. 103 is varied by taking for

Fig. 105.

one of the forms the symmetrical grouping, taken from the intercalation one composed of eight sections. The recurrence Fig. 104 is composed

Fig. 106.

of five elements, two of which are repeated three times, and one twice. In Fig. 105 we have three elements, one of them being repeated three times, and one twice.

Fig. 106 is composed of four elements; it can be considered as derived from a recurration of three elements, of which one would be repeated twice, and which would be cut in the intervals by the repetition of the fourth form.

By combining two and two and by intercalation of two sorts of arrangements, we obtain a great variety of ranges (Fig. 107).

Fig. 107.

The combination of an alternation of two forms, with a repetition of a group of three forms, gives Fig. 108.

128. The inevitable idea of number has a variable and relative importance to the mode of arrangement. In simple and uniform repetition the number is unlimited. In the alternation it is necessary to distinguish the parity and imparity, but only when the series is limited enough to be

Fig. 108.

appreciated at once. Alternation being correct, and finished, produces an uneven number of elements; it is then that a finished disposition is contained in the second class of dispositions. But if the series is sufficiently prolonged in order not to have to count the elements, the idea of number is nil. In intercalation, if the group of one and the most important element is composed of a great number, the inter-

calary form should be insufficient; it would be
repeated and the disposition be reduced to an
alternation of groups. Still, the idea of number
is secondary, and the essential idea which
intervenes in this case is the one of measure or
quantity. The number, very far from being an
active and direct principle of harmony, is, on the
contrary, quite subordinated to the æsthetical conditions of this harmony. In the periodical repetition it is necessary to maintain the whole period,
only, if these were too many, the continuation or
line becomes indefinite. In the recurration, which
is a disposition relatively complicated, the necessity
to maintain the symmetry is still more rigorous;
it is in this disposition that the ideas of number
expressly intervene, as well as symmetry, graduation, size, quantity, and æsthetically the ideas of
relief and colour, whose different particularities
show the different accents of the whole dispositions.

CHAPTER II.

RANGES.

129. The repetition and line of a form determine
ranges, where it is necessary to recognize two parts,
the headline and the composing forms. We distinguish two kinds of forms: the linear forms or
touches, and the superficial forms or plane figures.

Generally, considering the relative situation of
forms, they can be separated, continued or
crossed in the simple ranges, and articulated conjointly, or separated, continued or crossed in the

composite ranges. These ranges are divided into
two general classes : the marginal ranges, or those
only upon one side of the headline, and the diametrical ranges, whose parts co-exist on both sides of
the headline.

I. MARGINAL RANGES.

130. The marginal and lateral ranges have no
axes of symmetry. They contain two varieties :—

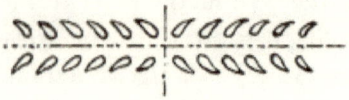

Fig. 109.

1st. The uneven ranges, equal to the lateral order,
or upon only one side, left or right. These ranges
are determined by the followed repetition of an

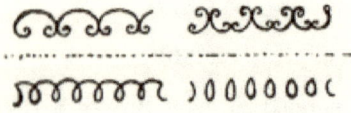

Fig. 110.

uneven form, and by the followed repetition of an
oblique even form. They can occupy four positions, viz. two situations, above or below the headline, and two directions, to the right or to the left
(Fig. 109).

2nd. The even or straight ranges, equal to the

vertical order, i.e., from top to bottom. These ranges are obtained by the followed repetition of a straight even form, and by the backward repetition of an oblique even form. These ranges can occupy two positions, the one on the top and the other under the headline, as in Fig. 110.

II. DIAMETRICAL RANGES.

131. The diametrical, collateral ranges have symmetrical axes which are confounded with the headline. They contain four varieties.

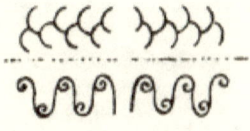

1st. Contradicted ranges. These are determined by the contradicted repetition of an uneven form, by the following or alternate repetition of an even form whose axis is horizontal, and by the constructed

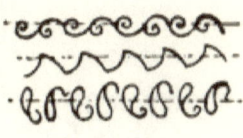

Fig. 111.

repetition of an oblique even form. These ranges can occupy four positions, which are reduced to two when the continuation is unlimited ; the difference of one position from the one which is symmetrical, or backward, consisting only in the moving of the form, the two symmetrical positions would be confounded by the removing, lengthwise, of the interval of a form (Fig. 111).

2nd. Diagonal ranges. These are obtained by the followed repetition of a diagonal form, a revolved, even or quartered form obliquely placed at its axis. These figures can occupy two positions (Fig. 112).

To ascertain whether a diametrical range is contradicted or diagonal, it is sufficient to change, in thought, one half on either the right or left. If one of them can be reduced so as to be symmetrical with the other, the range is contradicted, or it is diagonal.

3rd. Alternate ranges. These are obtained—

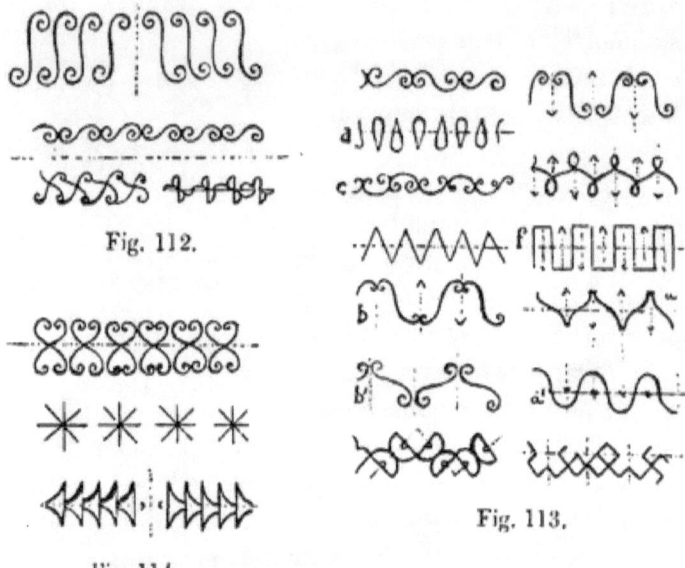

Fig. 112.

Fig. 113.

Fig. 114.

1. By the backward repetition of a diagonal form.

2. By the alternate repetition of a straight even form.

3. By the backward repetition of a revolved, uneven form, and a diagonal revolved form.

4. By the repetition, backward, of a quartered form and of a revolved form placed obliquely.

The alternate ranges can occupy only one position (Fig. 113).

4th. The opposite or straight ranges are—

1. Even, when the range is obtained by the following repetition of even form, or of a backward form whose symmetrical axes are confounded with the headline; these ranges can be directed on one side or the other.

2. Quartered, when the range is obtained by the followed or reciprocal repetition of a quartered form.

3. Centred, when the range is obtained by the followed repetition of a radiated form. The quartered or radiated ranges cannot occupy more than one position (Fig. 114).

CHAPTER III.

CONJUGATIONS OF RANGES.

132. By combining the positions, or different directions of the ranges two and two, we obtain conjugations similar to those of the conjugations of isolated forms.

Generally the conjugation is (1) followed, when the ranges are parallelly repeated; (2) returned, when the ranges are symmetrically repeated; (3) diagonal, when the ranges are diagonally inverted; and (4) contradicted, when the ranges are contradicted.

The conjugations are, for the reciprocal situation, (1) straight or directly corresponding; (2) alternated, when forms of a range are corresponding to the intervals of the other: and for reciprocal correlation (1) separated or distinctive; (2) continued, when ranges have a very small interval between them; (3) crossed, when the ranges are crossing one another by intersections. By making use of these different conditions for the ranges described above, we obtain enumerated and following figurations.

I. Marginal Ranges.

133. 1. Followed or uneven ranges. These two uneven determine four kinds of conjugations: followed, backward, diagonal and contradicted. We take, for example, range with simple or abrupt situation, to render the results more perceivable.

Followed Conjugations..

Fig. 115.

Straight and contiguous.

Fig. 117.

Alternate and separated.

Fig. 116.

Straight and crossed.

Fig. 118.

Alternate and crossed.

Backward Conjugations.

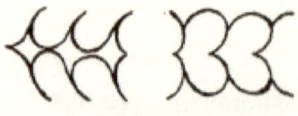

Fig. 119.

Straight and contiguous.

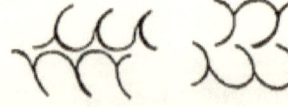

Fig. 121.

Alternate and separated.

Figs. 120, 125.

Straight and crossed.

Figs. 122, 126.

Alternate and crossed.

DIAGONAL CONJUGATIONS.

   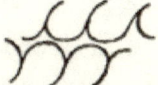

Fig. 123.

Straight and contiguous.

Fig. 124.

Alternate and separated.

CONTRADICTED CONJUGATIONS.

Fig. 127.

Straight and contiguous.

Fig. 129.

Alternate and separated.

Fig. 128.

Straight and crossed.

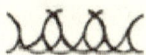

Fig. 130.

Alternate and crossed.

2. Straight or even ranges. The even ranges determine only the conjugations followed and backward.

FOLLOWED CONJUGATIONS.   BACKWARD CONJUGATIONS.

<table>
<tr><td>Fig. 131.</td><td>Fig. 134.</td><td>Fig. 137.</td></tr>
</table>

Fig. 131.	Fig. 134.	Fig. 137.
Straight and separated.		Straight and separated.

Fig. 132.

Alternate and contiguous.

Figs. 135, 138.

Alternate and contiguous.

Fig. 133.	Fig. 136.	Fig. 139.
Alternate and crossed.		Crossed.

II. DIAMETRICAL RANGES.

134. 1. *Uneven and contradicted ranges.* For using up all the possible conjugations, it is necessary, firstly, to consider the ranges having an even number of forms, which are regular and with contradicted symmetry (Fig. 140) ; the ranges having an uneven number of forms, which are uneven or unsymmetrical (Fig. 141) ; and the ranges having an incomplete number of forms, and finished on one extremity by dropping a form, which are equally uneven or unsymmetrical (Fig. 142) : and, secondly, to conjugate the two ranges by a turning down, straight, alternate and half alternate. The juxtaposition is a straight, when the forms are directly responding ; it is alternate, when the juxtaposition is produced by the retardment of a form, one of the labels being changed in the longitudinal way of an interval of form; and it is half alternate when the juxtaposition is immediate, between the straight and alternate juxtaposition, the moving being reduced to the fraction of interval.

Fig. 140.   Fig. 141.   Fig. 142.

Hence, and taking for example the linear and geometrical figuration, or composed arcs of a circle, we obtain six conjugations : A (Fig. 143), B (Fig. 144), c (Fig. 145), D (Fig. 146), E (Fig. 147), and F (Fig. 148).

The conjugation A is obtained by the lateral conjugations of two even foliages; by alternate con-

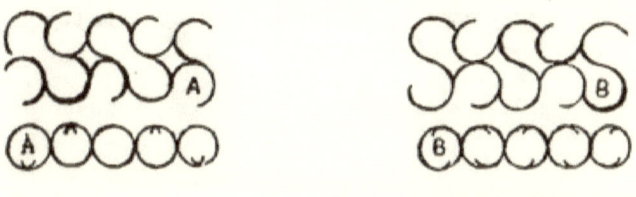

Fig. 143.   Fig. 144.

jugation of two diagonal even ranges; by the straight conjugation of two diagonal even foliages; by the half alternate conjugation of two diagonal incomplete ranges; and by the straight conjugation of two uneven ranges with longitudinal turnings.

The conjugation B is obtained by the conjugation of two ranges with lateral turning; by the conjuga-

L

tion of two uneven ranges with alternate juxtaposition; by the conjugation of two even ranges
with straight juxtaposition; by the conjugation of
two incomplete ranges with half-alternate juxta-

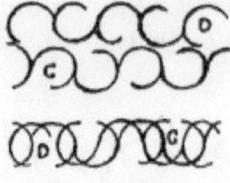

position; by even and half
alternate ranges with straight
juxtaposition; by diagonal
conjugation of two straight
uneven ranges and of two
alternate even ranges; and
by the conjugation, backward

Figs. 145, 146.

and longitudinal, of two uneven ranges, even and
incomplete with alternate juxtaposition.

The conjugation c is obtained by the lateral back-

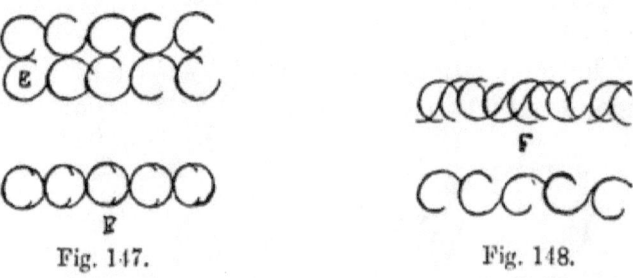

Fig. 147.   Fig. 148.

ward conjugation of two half-alternate, even and
alternate incomplete ranges; and by the diagonal
conjugation of half alternate uneven and straight
incomplete ranges.

The conjugation D is obtained by the lateral
and straight backward juxtaposition of incomplete
ranges; by the diagonal and half alternate juxtaposition of incomplete ranges; by diagonal and
alternate juxtaposition of incompelte ranges; and

by the juxtaposition of a longitudinal and half alternate turning of uneven ranges.

The conjugation E is obtained by straight juxtaposition with a lateral turning of uneven ranges; and by the straight juxtaposition with longitudinal turning of incomplete ranges.

The conjugation F is obtained by juxtaposition with lateral and half-alternate turning of uneven ranges; by followed and half-alternate juxtaposition of uneven, even and incomplete ranges; and by juxtaposition with longitudinal and half-alternate turning of even and incomplete ranges.

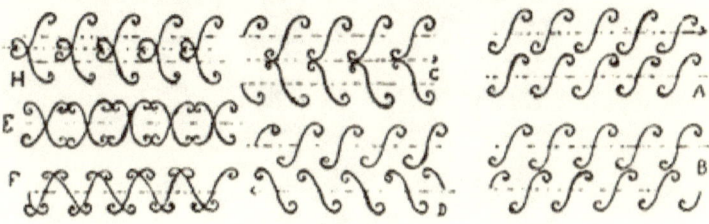

Fig. 149.

2. *Diagonal ranges.* Two diagonal ranges determine the followed and backward conjugations, which can be straight, alternate and half alternate, distinct, contiguous or crossed. By conjugating the diagonal ranges obtained from a revolved or quartered form, we should obtain the same ranges, but particularly the backward conjugation; alternate or intersected, determines an alternate diametrical range. Two diagonal ranges obtained from a simple diagonal form being conjugated backward and of half-alternate juxtaposition determine an even marginal range.

3. *Alternate ranges.* The alternate ranges being

conjugated, determine the followed and backward conjugations, whose correlation of position can be straight, half alternate, or alternate. The half alternate crossed conjugation is alternate diametrical, ɢ (Fig. 150). The alternate followed conjugation and the straight and backward conjugations are quartered diametrical, ᴅ.

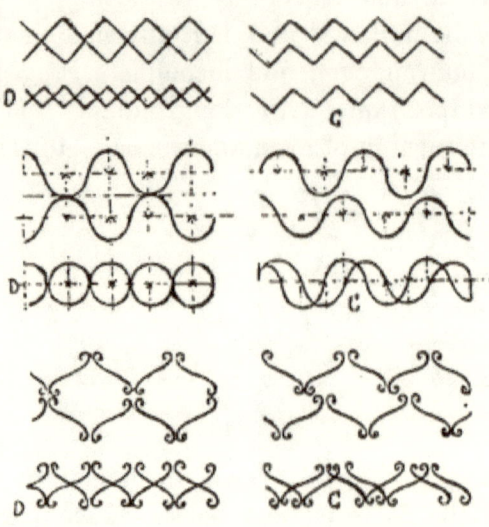

Fig. 150.

4. *Opposite or straight ranges.* Two even diametrical ranges determine the followed and backward conjugations. The followed conjugation is even diametrical if the juxtaposition is straight; it is contradicted, diametrically, if the juxtaposition is alternate. The backward conjugation is diagonal diametrically, if the juxtaposition is straight, half alternate or alternate (Fig. 151).

Two quartered diametrical ranges, or centred

diametrical, determine the followed conjugation, which is straight diametrical if the juxtaposition is straight, and alternate diametrical if the juxtaposition is alternate (Fig. 151).

135. The preceding considerations refer to every

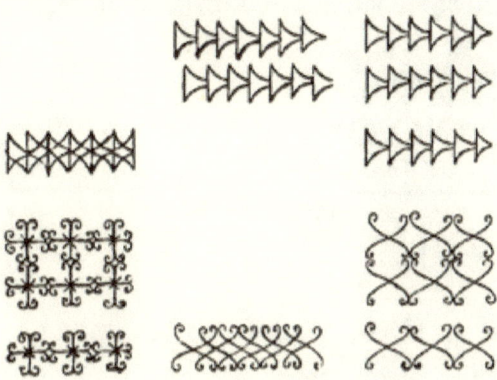

Fig. 151.

sort of ranges, but it is very useful to consider particularly the ranges determined by articulation of forms with a real or imaginary headline. Two simple ranges placed in series can be repeated or doubled in the same way or in opposition, and by straight or alternate juxtaposition, and the following diagrams are obtained :—

I. Unilateral Dispositions.

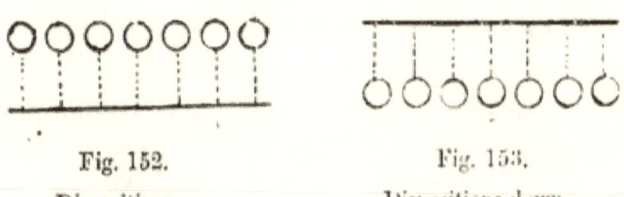

Fig. 152.

Dispositions up.

Fig. 153.

Dispositions down.

These two dispositions, simple and varied by inclination of forms on the headline, being doubled in the same way by superposition or intercalation, determine the following dispositions :—

1. Straight superposed disposition.
2. Alternate superposed disposition.

Fig. 154.   Fig. 155.

II. Collateral Dispositions.

The two dispositions, up and down, being conjugated backward, determine the four distichous ranges :—

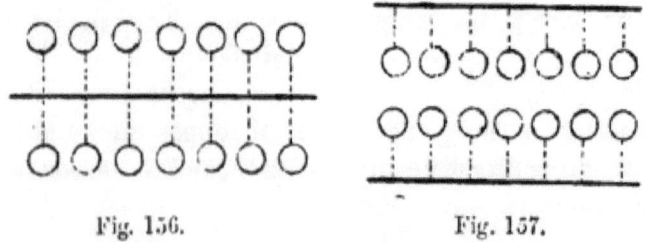

Fig. 156.   Fig. 157.

1. Straight opposite disposition (Fig. 156).
2. Straight face to face disposition (Fig. 157).
3. Alternate opposite disposition (Fig. 158).
4. Alternate face to face disposition (Fig. 159).

These diagrams, simply figurative, have a real or imaginary axis. If the axis is real, its form can be varied; it is then a delineation. We have supposed hitherto that the ranges are uniform and composed with only one form; it is necessary now to examine

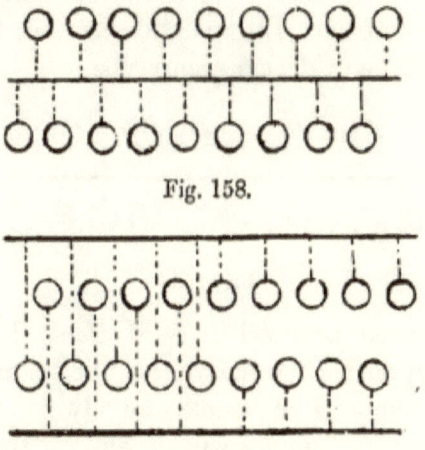

Fig. 158.

Fig. 159.

the different positions given by the conjugation of varied forms, i.e., composed of different forms.

CHAPTER IV.

VARIED DISPOSITIONS.

136. The five fundamental rhythms, combined two and two, and referred to in the six preceding diagrams, determine ninety dispositions.

To avoid dwelling too long on this paragraph, we shall limit our examination by the two following conditions: by taking for example only the different combinations and ranges having at most three different forms, and referring to the construction in only one diagram of the straight, opposite disposition. We shall thus get the following arrange-

ments:—Repetition alternation, Repetition intercalation, Repetition period, Repetition recurrence;
Alternation intercalation, Alternation period, Alternation recurrence; Intercalation period, Intercalation recurrence; Period recurrence.

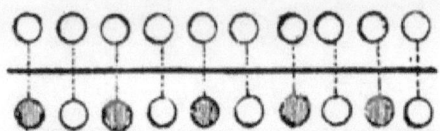

Fig. 160.

The degree of excellence of these dispositions
depends upon their relative simplicity, and, seeing
that the combined elements are always the same,
the degree of simplicity will depend on the size and
extent of period and of symmetrical correlation
which is peculiar to it.

1. Repetition alternation has one period of two
forms. This disposition is unique. (Fig. 160.)

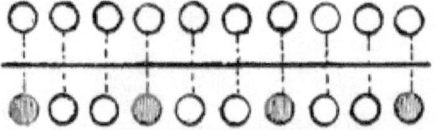

Fig. 161.

2. Repetition intercalation. If the intercalation
has a period of 3, 4, 5 . . . . elements, disposition has also a period of 3, 4, 5 . . . . elements.
This disposition can be varied, so as to give two
different forms. (Fig. 161.)

3. Repetition period has the same period as
periodical ranges. By transposition, two forms
are obtained. (Fig. 162.)

4. Repetition recurrence has a period of four elements, the recurrence having three forms; and

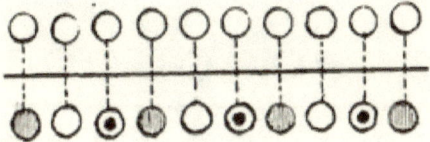

Fig. 162.

a period of eight elements, recurrence being of four forms. Transposition gives three varieties. (Fig. 163.)

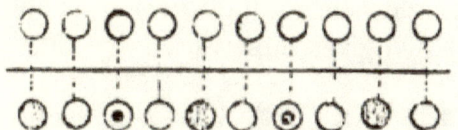

Fig. 163.

5. Alternation intercalation has a period of six elements if intercalation has a period of three elements; and a period of four elements if the period of intercalence is of four elements. Generally, intercalations of uneven period determine

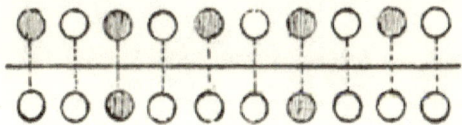

Fig. 164.

in the dispositions a period of double number; and intercalences of even period a period of the same number. By transposition two varieties are obtained. (Fig. 164.)

6. Alternation period. If the period of range is

even, period of disposition is of the same number ; if it is uneven, a period of disposition is of a double number.   Removal gives two varieties.   (Fig. 165.)

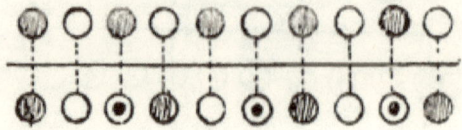

Fig. 165.

7. Alternation recurrence has a period of four elements if recurrence is of three forms, and a period of eight elements if recurrence is of four forms. Transposition gives three varieties.   (Fig. 166.)

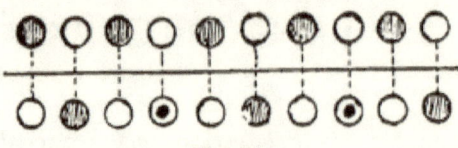

Fig. 166.

8. Intercalation period gives a period of three forms if the cuttings of the two composing ranges are of three forms.   Generally, if the periods are of the same number, they determine in disposition an

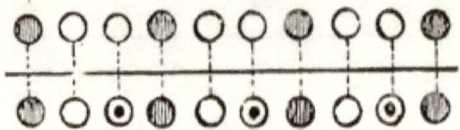

Fig. 167.

equal period ; if there is discordance, the composed period is of 12, 18. . . . &c., elements.   By transposing the period, six forms are produced from the dispositions.   (Fig. 167.)

9. Intercalation recurrence produces a period of four elements if intercalation has a period of four elements; and a period of twelve if the period of intercalation is only of three elements. Transposition gives three varieties. (Fig. 168.)

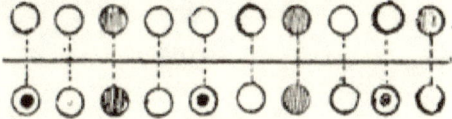

Fig. 168.

10. Period recurrence has a cutting of four elements if the period is of four forms ; and a cutting of twelve elements if the period is of three forms. Transposition gives three varieties. (Fig. 169.)

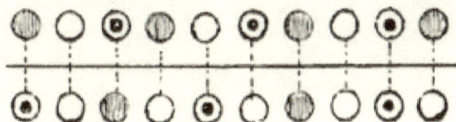

Fig. 169.

The dispositions repetition alternation, repetition recurrence, and alternation recurrence are the simplest and most regular.

MULTIPLIED DISPOSITIONS.

ALL the possible diagrams of plane co-ordinated dispositions have for their essential basis the theory of networks.

CHAPTER I.

NETWORKS.

137. When points are disposed in a regular manner on the whole extent of a plan, they form immediately real or imaginary lines which bind them together in finished or indefinite lines, crossed, and regularly bound, and determining points regularly disposed. These points and lines determine networks. Moreover, the points determine distances, and these distances are straight lines, and together they determine polygonal spaces, which are segments of the plan.

For determining networks we start indifferently from the distribution of points, from the crossing of the straight lines, or from the repetition of polygons. The networks can be seen under a triple point of view : 1st, points ; 2nd, lines ; 3rd, meshes.

I. Generation by Points.

138. A series of points placed in line can be put near another similar series in two ways. The first, where the points are directly corresponding and

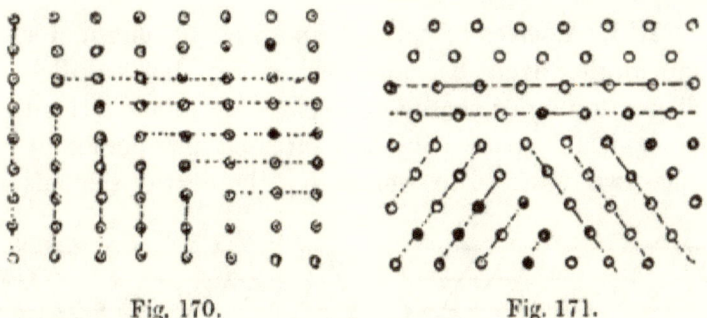

Fig. 170.   Fig. 171.

following lines perpendicular to the headline; this is the straight repetition (Fig. 170). The second, where points change places on the right or left side and correspond to the intervals; this is the

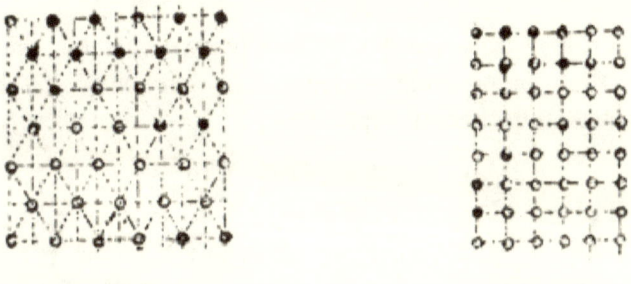

Fig. 172.   Fig. 173.

alternate repetition (Fig. 171). Distances can be smaller, or of the same size, or even greater than the intervals of points on the range. From the regular seed-bed of these points come horizontal, vertical or oblique lines (Figs. 172, 173); or four systems for the straight repetition, two rectangular and diagonal systems, and six systems for the alternate repetition, as well as three principal triangular systems, and three diagonal systems, also triangular.

II. Generation by Lines.

139. A series of parallel lines can be cut under an angle, even by another series also parallel. The crossing of these two series determines a third one, which passes by the intersecting points of the two first. But generally, these two systems

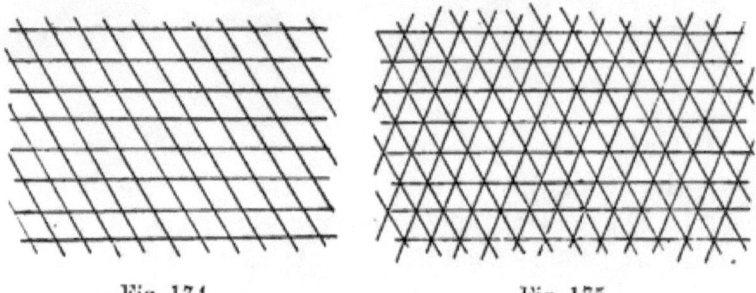

Fig. 174.   Fig. 175.

of crossed lines are cut by one or two systems, each of them being independent of or correlative to the others, together or separately.

III. Generation by Figures.

Polygonal figures determine four classes of networks: 1, quarternary networks; 2, senary networks; 3, disjoined networks; 4, composed polygonal networks.

Class I.   Quarternary Networks.

140. These networks are determined by parallelograms, rectangles, lozenges, and the square, placed side by side in perfect coincidence.

1. Parallelograms. The centres of figures determine the same network; the diagonal lines determine a network equally parallelogramic, but not of the same form (Fig. 176).

2. Lozenges. The centres of figures determine

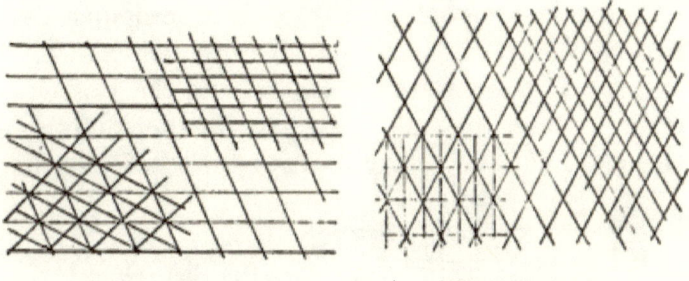

Fig. 176.   Figs. 177, 179.

the same network; the diagonal lines determine a rectangular network (Fig. 177).

3. Rectangles. The centres of figures determine

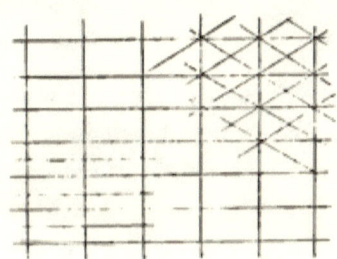

Fig. 178.

the same network; the diagonal lines determine the lozenge network (Fig. 178).

4. Square. The centres of figures determine the same network, and the diagonal line determines a squared network (Fig. 180).

All these networks can be varied in every degree,

following the uniform separations; and varied in every system, following the similitude or diversity of separations from one system to the other.

CLASS II. SENARY NETWORKS.

141. All the networks of this class are determined by three systems of lines, the third crossing the two first ones in their intersecting points. The elementary meshes are the scalens and the isosceles triangles and trigons. In all the networks of this class it is necessary to distinguish between the

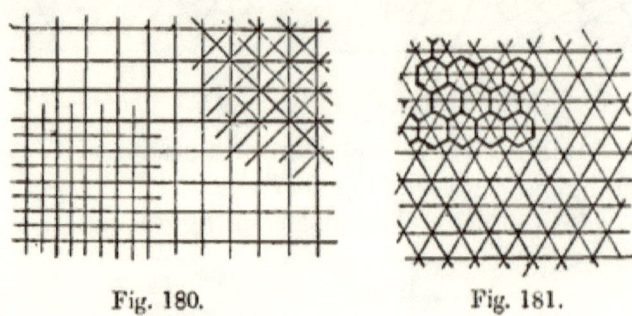

Fig. 180.   Fig. 181.

isosceles networks, whose mesh is isosceles, and the diagonal networks, whose mesh is trigonal.

These first two classes of networks are reduced to three principal systems :—

1. Trigonal network, with trigonal mesh (Fig. 181).

2. Squared network, with squared mesh (Fig. 180).

3. Isosceles and conjugated networks, with rhombical mesh, or lozenge with rectangular mesh (Fig. 179).

CLASS III. DISJOINED NETWORK.

142. If we consider only the joining of polygons,

we obtain, with the polygonal figures, at first the networks described above, and then the disjoined networks.

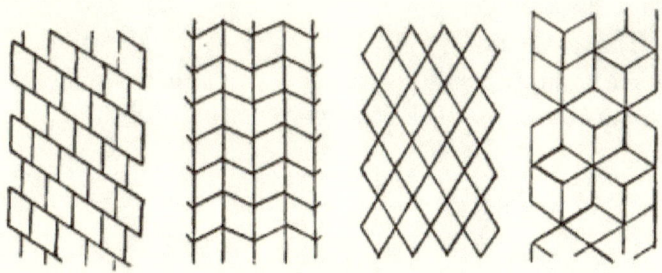

Fig. 182.

1. Lozenges. Assembled lozenges give, by followed juxtaposition, the ordinary network where the lozenges have all the same direction; by juxtaposition backwards, the lozenges have two direc-

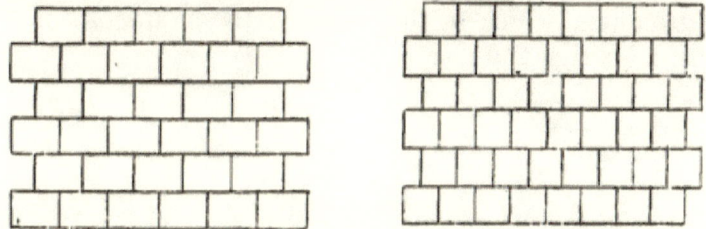

Fig. 183.

tions; by centred juxtaposition, the network has three directions, a varied network by combination of three first juxtapositions; and also the juxtaposition determined by three lozenges assembled in point and in side (Fig. 182).

2. Rectangles and Squares. The rectangle and the square determine the followed and rough juxtaposition (Fig. 183).

M

3. Trigon. The juxtaposition of trigons is followed and turned, and contains a particular case and varied groupings of the different networks of trigonal lozenge.

If we do not pay attention to this condition of perfect juxtaposition we get other arrangements which depend on the relative position of figures.

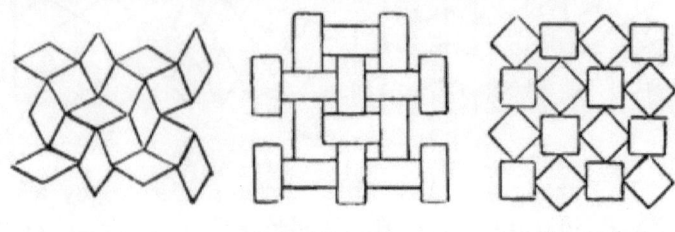

Fig. 184.   Fig. 185.   Fig. 186.

1. Lozenges. The lozenges placed alternately one way and the other determine the followed grouping, which is the same as the juxtaposition of a lozenge and a square (Fig. 184).

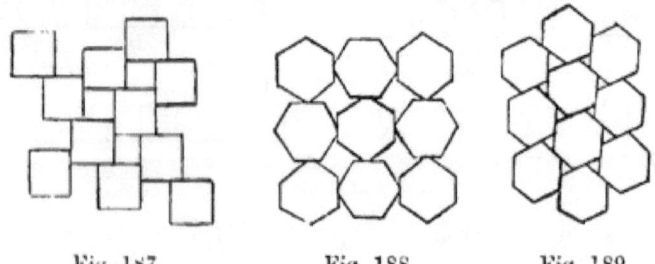

Fig. 187.   Fig. 188.   Fig. 189.

2. Rectangles. Two rectangles whose positions are contradicted give the combined network of a rectangle and a square assembled (Fig. 185).

3. Square. The two positions of a square, placed on the point or side, determine the contradicted grouping (Fig. 186). The followed positions of

the square, with removal, give a figuration equal to that of grouping two conjugated squares (Fig. 187).

4. Hexagon. Like the square, this form has the contradicted grouping and the followed grouping, which coincide with the juxtaposition of a trigon and a hexagon (Figs. 188 and 189).

5. Trigon. The figuration obtained equals the juxtaposition of the trigon with the hexagon, conjugated, i.e., one being given, the other is not arbitrary (Fig. 190).

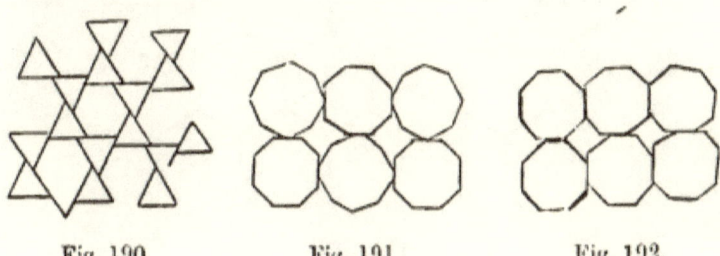

Fig. 190.   Fig. 191.   Fig. 192.

6. Octagon. We obtain figures analagous to the preceding (Figs. 191 and 192).

All the forms of juxtaposition that we can refer to polygons are very simple.

CLASS IV. COMPOSED POLYGONAL NETWORKS.

143. The condition of perfect juxtaposition, with only one polygon, is produced by a hexagon whose juxtaposition follows the position, disjoined for lines of network, and polygonal because the figure has more than four sides. The network of a hexagon and succeeding networks (Figs. 193, 194, 195, 196,

197) are constructed on the type of the trigonal network. The network of the square and such like
(Figs. 198, 199, 200, 201, 202) are constructed from
the type of square network.

1. Hexagon (Fig. 193).

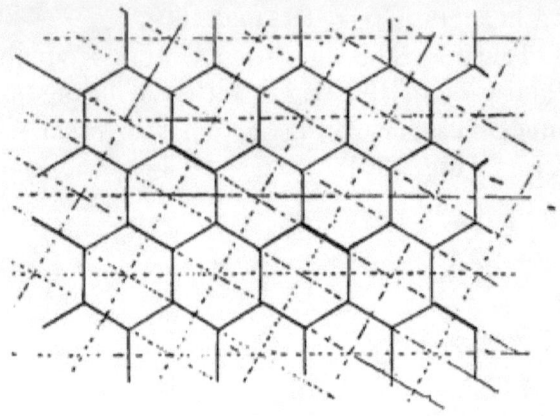

Fig. 193.

2. Hexagon and Trigon (Fig. 194).

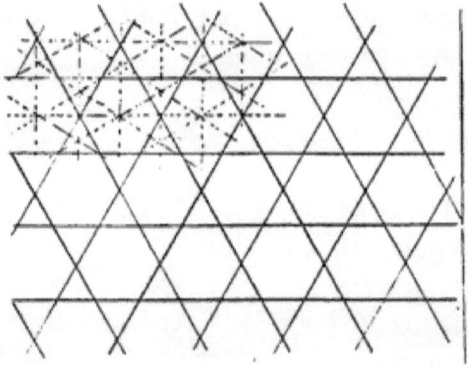

Fig. 194.

3. Dodecagon and Trigon (Fig. 195).
4. Hexagon, Square, and Trigon (Fig. 196).

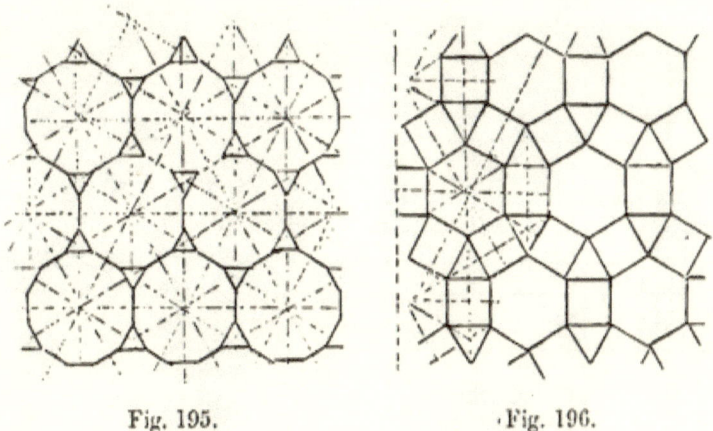

Fig. 195.   ·Fig. 196.

5. Dodecagon, Hexagon and Square (Fig. 197).

1. Octagon and Square (Fig. 198).

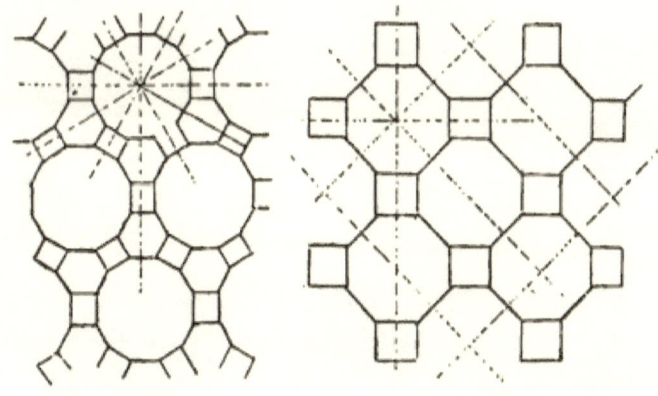

Fig. 197.   Fig. 198.

2. Rectangles and Squares (Fig. 199).
3. Lozenges and Rectangles (Fig. 200).

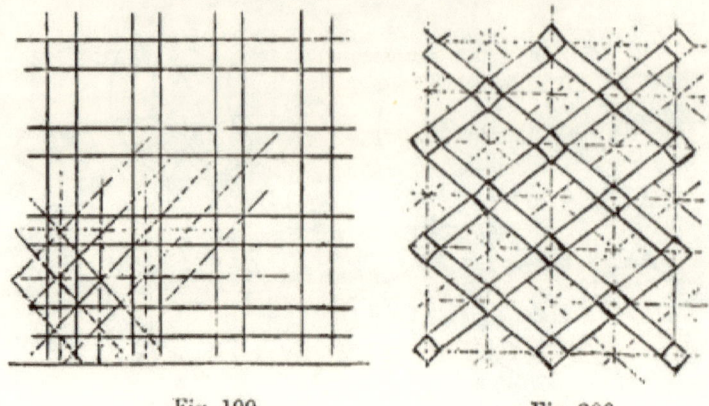

Fig. 199.   Fig. 200.

4. Lozenge and Square (Fig. 201).
5. Parallelogram and Squares (Fig. 202).

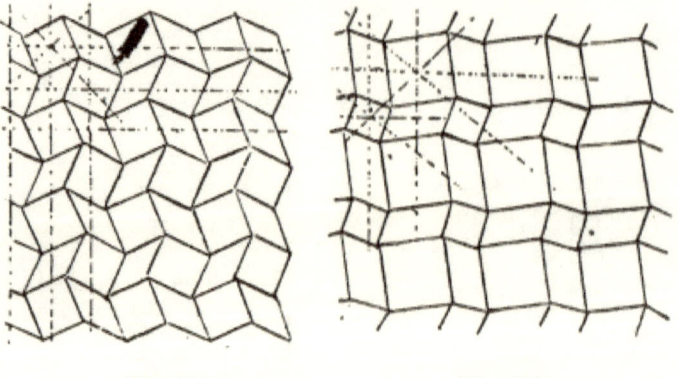

Fig. 201.   Fig. 202.

All the regular figures which by their juxtaposition determine their networks have their sides
equal. If we consider only the summits, all the
points are at equal distances from the homologous

or corresponding parts, and equally distributed upon the whole extent of a plan.

CHAPTER II.

GROUPING OF CIRCLES.

144. The two or three fundamental groupings are the following : three tangent circles, following the trigon ; four, following the square, and four following the trigonal lozenge.

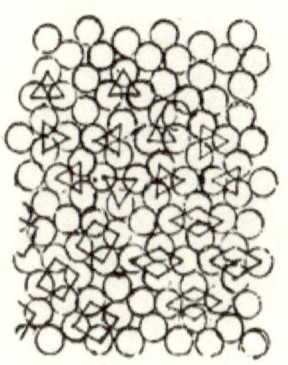

Fig. 203.

The complete grouping in number is produced by the agglomeration of elementary groupings, trigon, square, lozenge, rectangle, hexagon and dodecagon; or it may also be obtained by distributing the circles following the polygonal networks derived from these figures.

I. Lozenge, Trigon, Square.

With a little attention, we can recognize, in this map, different groupings: a grouping of four circles in the square, under two positions ; a grouping of four circles in the lozenge, under two positions ; a grouping of three circles in the trigon, under four positions ; and a grouping of two circles under two positions. These last are discontinued, whilst the integral groupings, on the contrary, are continued, following the complete network of lozenge and square (Fig. 203).

II. LOZENGE, LOZENGE-RECTANGLE.

The example chosen contains a lozenge of four trigons, a lozenge of two trigons, and a rectangle of two squares. This disposition has two assembled groups of lozenges composed of nine tangent circles, or even groups of lozenges and hexagons (Fig. 204).

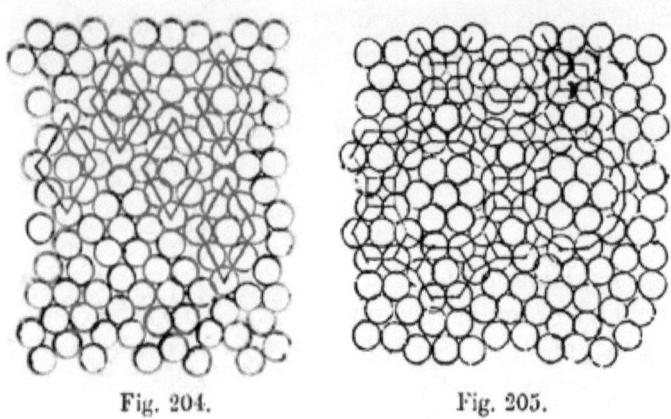

Fig. 204.   Fig. 205.

III. HEXAGON, SQUARE, TRIGON.

This disposition results from a grouping of hexagons composed of seven circles (Fig. 205).

IV. HEXAGON, TRIGON, RECTANGLE.

This disposition results from hexagons composed of nineteen circles.

Hexagons, trigons and squares could be composed of a number of circles (Fig. 206).

V. HEXAGON, SQUARE, LOZENGE.

This disposition is the result of a varied grouping composed of three circles in trigon and seven circles in a hexagon.

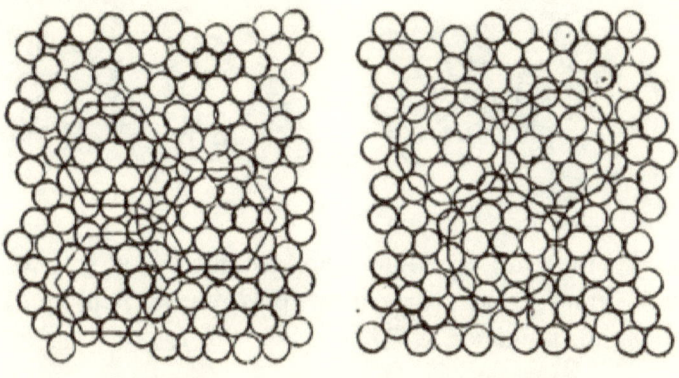

Fig. 206.   Fig. 207.

VI. DODECAGON AND TRIGON.

The agglomeration results from a complete grouping of hexagon, rectangle, trigon, or grouping of hexagons and two circles in line, &c. (Fig. 207).

VII. DODECAGON, HEXAGON, SQUARE.
VIII. DODECAGON, HEXAGON, SQUARE, TRIGON.

This agglomeration results from a complex and discontinued grouping of dodecagon and trigon (Fig. 208).

As to the other polygonal networks, we can add equal circumferences following the line of network or the sides of polygons; but such arrangements leave variable inter-

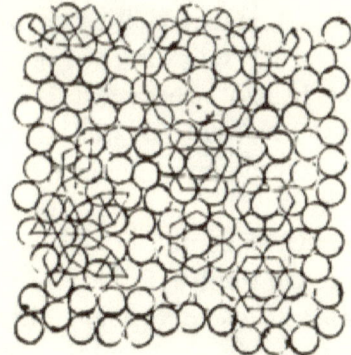

Fig. 208.

vals. Following the figure the interval is smaller than the circle of grouping for the trigon, square,

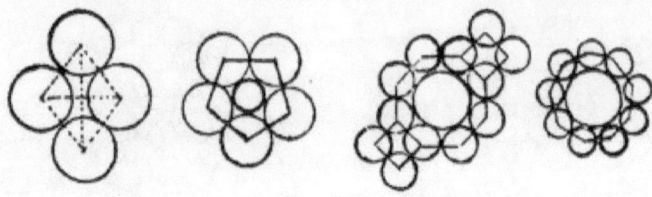

Fig. 209.

and pentagon ; equal for the hexagon ; and greater and greater for every other polygon, as heptagon, octagon, nonagon, &c., &c. (Fig. 209).

CHAPTER III.

REPARTITIONS.

145. The theoretical and general notions once established, the application to ornamentation soon results. This application is explained in the three following categories.

1. Ornamentation by separated forms and placed with order, following the points of networks. These repartitions are varied according to the different rhythms of ranges, and determined by isolated opposition of forms in each of the head-points of the network.

2. Ornamentation by rays or separated and directed ranges, following the harmonical lines of plan, which determine three systems of rays or ranges—the horizontal, vertical, and oblique or diagonal.

3. Ornamentation by squares and stripes crossed. These crossings are determined by combination two and two, three and three, four and four, of the different systems of stripes, horizontal, vertical, or oblique.

The squares, ranges or rows, and separated forms being combined in every possible manner, determine maps of ornament which we call tapestry.

146. At first the points are placed at random and cover the surfaces; these surfaces are destitute of order and represent Nature. Each of them having

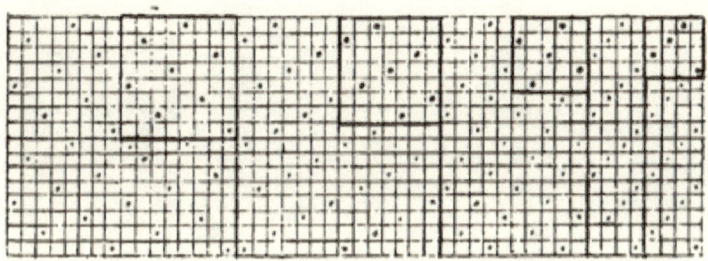

Fig. 210.

been under man's manipulation, it is then exclusively decorative. Stuffs or kinds of decoration in colour imitate them; they dispose or separate the seed-bed by being in contact with diagrams of disposition. The diagrams are of several kinds: here are the principal.

1. Diagrams determined by square network, rectangles or lozenges. The points are placed following the points of network, or following the centres of meshes, every touch disappearing.

2. Diagrams determined by networks disjoined,

though the repartition may be at the meeting-point of lines, or at the centre of meshes.

3. Diagrams determined by polygonal networks. The distribution of points is one of the features of network or centres of meshes. All the repartitions are alike. But if we employ 2, 3, 4, 5 . . . forms, we get varied or composed repartitions. The diagrams are networks, and we can recognize in them, as well as in uniform repartitions, ranges or squares, i.e., co-ordinated repartitions, in the two dimensions

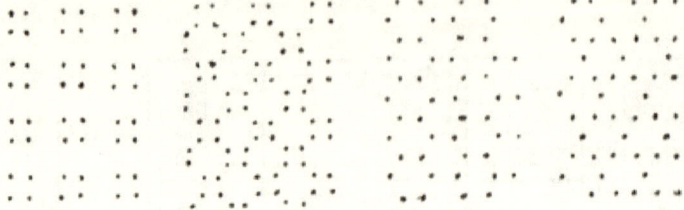

Fig. 211.

of extent. We will give a few examples of repartitions of 2, 3, 4, 5 forms.

I. Repartition of Two Forms.

147. 1. Following the square network, and the varieties of this network. The figure 212 offers only uniform ranges in the vertical, and alternations in the horizontal way. The figure 213 has uniform ranges of alternate succession, following horizontal and vertical lines, and alternations following diagonal lines; this disposition is poised. The figure 214 is a square network cut by another network whose mesh is a square inscribed in the first one. The horizontal and vertical

ranges are alternations in the great network, and
the diagonal ranges are repartitions.

2. Following the trellis-work and the varieties of
this network. The figure 215 is a series of repar-

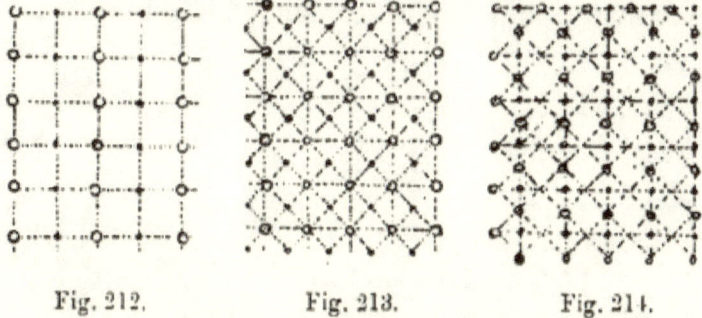

Fig. 212.   Fig. 213.   Fig. 214.

titions or ranges, the oblique ranges being uniform
and the horizontal alternations. The figure 216
is a figured or poised repartition, and it is composed of hexagonal repartition centred by trigonal

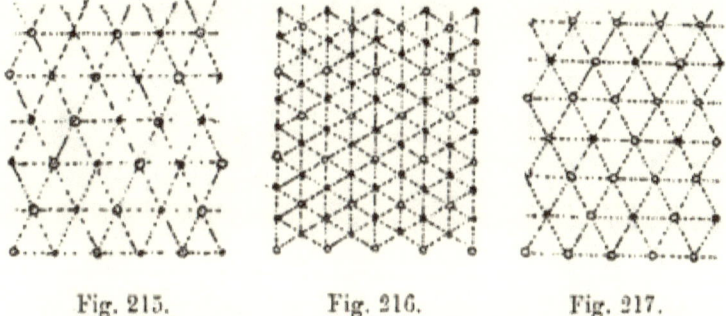

Fig. 215.   Fig. 216.   Fig. 217.

repartition. All the rows are intercalations two and
one. The figure 217 is a repartition figured and
composed of a polygonal, trigonal, and hexagonal
network, whose hexagons are centred by a trigonal
network; the ranges are alternately repetitions

and alternations. These three figures are analogous.

II. Repartition of Three Forms.

148. 1. Following the square network and varieties of this network. The figure 218 is a

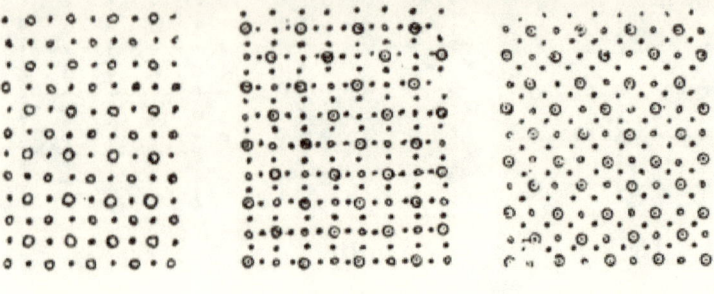

Fig. 218.   Fig. 219.   Fig. 220.

figured repartition whose rows are all alternations, or three alternations resulting from the combina-

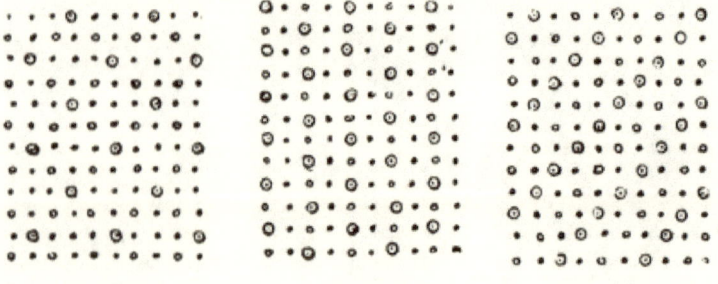

Fig. 221.   Fig. 222.   Fig. 223.

tions two and two of three forms. The figure 219 is a figured repartition of which all rows are recurrences. The figure 220 is a figured repartition, all of whose straight rows are alternations. The intercalary and alternate ranges are repetitions. The figure 221 is a figured repartition which has all its

straight rows alternately one alternation and one
intercalation, its diagonal rows being alternately one
alternation and one repetition. The figure 222 is a
repartition by series or ranges, the vertical alternately repetitions and alternations, and the hori-

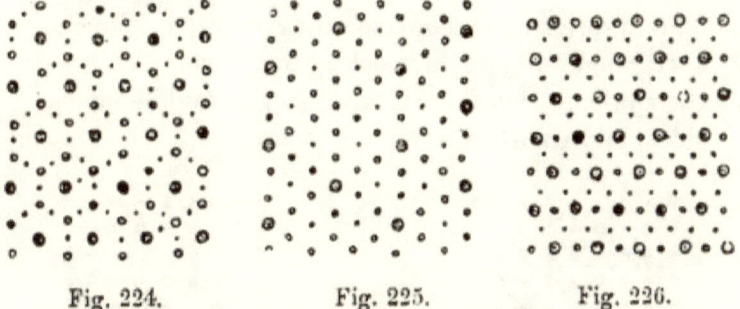

Fig. 224.   Fig. 225.   Fig. 226.

zontal recurrences. The figure 223 is a repartition by series. All the horizontal and vertical
rows are recurrences, the diagonal ranges being

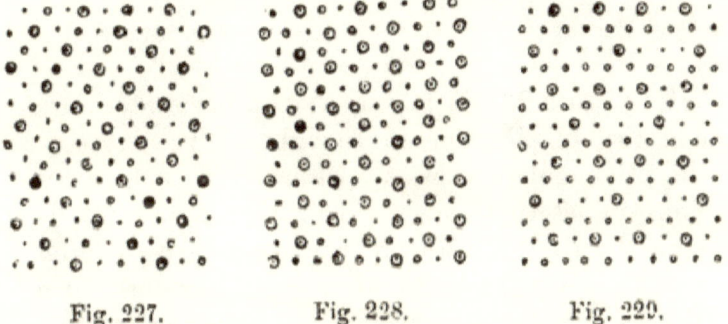

Fig. 227.   Fig. 228.   Fig. 229.

from left to right alternately alternations and repetitions, and from right to left repetitions.

2. Following the trellis network, and varieties of
this network. The figure 224 is a figured discontinued repartition, determined by the summits, the

middle of the sides and the centre of the mesh.   The
figure 225 is a figured repartition, trigonal ; a form
is placed at the summits, another at the centres, and
the third at the triparted sub-division of sides.   All
the rows are intercalations.   The figure 226 is in
series ; all the rows are alternations or repetitions.
This figure can be considered as ordered following
a trigonal lozenge network ;   one form placed at
the summits, another following the middle of the
sides, and the third at the centres.

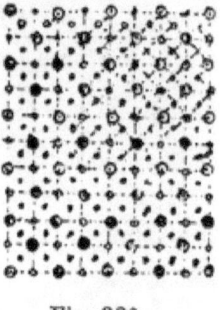
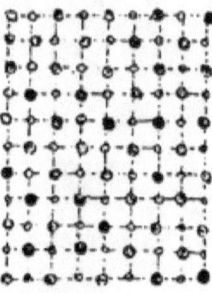

Fig. 230.   Fig. 231.

The figure 227 is in series ; the horizontal rows
and the oblique from left to right are recurrences ;
the other oblique rows are alternations and repetitions in alternate succession.   The figure 228 is
figured ; all the rows are periods or a repartition
following the summits of trigonal network, then
alternately from   trigon   to   adjacent   trigon,
and are the repetitions of the second and third
form.   The figure 229 has its horizontal rows
alternately alternations, repetitions and periods of
three and one ; the oblique rows are alternately
alternations and recurrences.

III. Repartition of Four Forms.

149. 1. Following the square and correlative networks. The figure 230 and the figure 218 have intercalated a fourth form, following the centres; all the diagonal rows, in 230, are recurrences. The figure 231 is a mixed repartition with quartered symmetry; all the straight rows are alternations.

2. Following the trellis and correlative networks. The figure 232 is analogous to the figure 231, all the rows being alternations. The figure 233 is a figured

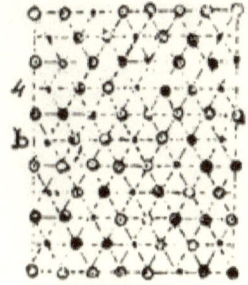

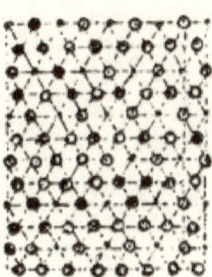

Fig. 232.   Fig. 233.

repartition; all the rows are alternately periods and varied seriations resulting from the intercalation of two series, one periodical, the other repeated.

IV. Repartitions of Five Forms.

150. The figure 234 is a figured repartition. The horizontal and vertical rows are following in order of recurrence, the rows being of three kinds: recurrence of three forms, repetition of one form, and alternation of two forms. The diagonal rows are alternately one recurrence of four forms, one alter-

N

nation of two forms, and one recurrence of three

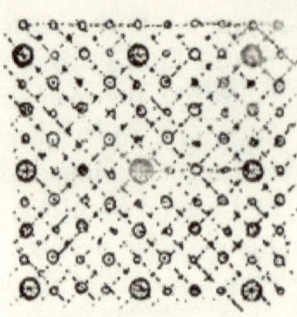

Fig. 234.

forms. These diagonal rows are disposed, like the straight rows, in recurrent succession.

The figures 235, 236, 237, 238, 239 are squares, whose abatements should determine other repartitions; thus, the figure 235 is the square of the figure 230.

Fig. 235.

Fig. 236.

Fig. 237

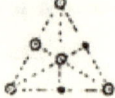

Fig. 238.

Fig. 239.

THIRD SECTION.

AGGLOMERATED DISPOSITIONS.

151. This order of dispositions contains :—

1. Hypothetical or real agglomeration of atoms, molecules or particles, which constitute the material in its different specifications.

2. Composition of organized tissues or materials of organism, which constitute the tissues or bodies of the animal and vegetable world.

3. Natural or artificial agglomerations, determined by joining distinct elements, without any bond other than the external cause constituting them.

From the indefinite agglomerations to the definite or figured dispositions there is a continued passage preserved by indefinite or uncertain groupings. Hence the three terms: agglomerations which are indefinite; uncertain groupings; and definite dispositions. These last form a distinct class, supplying itself by the principle of unity or individuality which separates and determines the individuals which compose it. This principle of unity is organization in the case of living beings, and composition or æsthetical elaboration in the case of artistic works.

Uncertain groupings are portions or limitations

of indefinite agglomeration, and constitute a form-envelope independent of the inner arrangement. These limitations are analogous to material forms; and both are specified by the conditions of an inner symmetry and regularity correlative to the form-envelope. The form-envelopes being brought into contact with the abstract and geometrical forms essential to the uncertain groupings and definite dispositions, a tree can present a pyramidal configuration as well as a heap of sand or a constructed pyramid.

FIGURATIVE DISPOSITIONS.

152. All these dispositions have a finished diagram, and are divided into three classes:—

1. Limited portions of ranges. These portions are limited or finished by the intervention of one or several of the following conditions, viz.: Headline, figure-envelope, symmetry, and declination.

2. Limited portions of repartitions in surface, or groupings limited by the two conditions of figure-envelope and symmetry.

3. Figurative dispositions which admit in their construction the intervention of the following particulars: Diagram of construction determining the type, headline, figure-envelope, symmetry, declination, and variation.

CHAPTER I.

LINEAR DISPOSITIONS.

153. A portion, whether it be of marginal or diametrical range, does not constitute a figurative disposition. Its construction is one of range, and its form-envelope an equal portion of indefinite form of the range; it is necessary to have other restrictive and special conditions in order that this portion may become a disposition really finished, and that implicates unity or plan of dispositions. The ranges possess :—

1. One limited headline, straight, arc, curved, handle, &c.
2. One finished form-envelope.
3. A limited number of elements.
4. One declined succession.

The conditions of symmetry and conjugation do not change in ranges and limited dispositions. The straight headline, being limited, has two extremities, one virtual middle or centre of figure, and symmetrical sub-divisions on both sides of the middle; or, considering the headline in only one sense, the straight line has a beginning and end, the sub-division of interval being graduated in progression, and by the declination of forms which follow. The straight line, in only one direction, takes the parity on both sides, the headline becoming a symmetrical axis. The straight, in two directions, takes the even or quartered symmetry, the transversal and longitudinal axes being cut at the centre of the figure.

Form-envelope, rectilinear or curvilinear, according to the character of declination, may be an uneven, even or quartered, diagonal or contradicted figure. It may also be open, or angular, or oval, or closed, as well as rectangular, lozenge, &c.

The horizontal linear order requires symmetry, i.e., a middle, which is known, and likeness on both sides. At a front view, it is essential that there be a middle, where what is on both sides can be brought into contact. According to the character of disposition, which can be an ornament or a monumental ordinance, this linear order is capable of marginal, quartered or diametrical symmetry; further, the extremities are more or less visible. The last diagram of this order is the straight line limited, with quartered symmetry, where it is necessary to distinguish one longitudinal axis, one transversal passing by the middle, and the two extremities.

In vertical order, or from the top to the bottom, it is on the base that they come in contact with the disposition, which is the starting-point. There is only one regular succession from the top to the bottom. The vertical linear order has then for its diagram a straight line, considered only as following a symmetrical and uniform direction on both sides of its length.

Generally and independently of all localization of space, these two kinds of order can be thus named :—

1. Symmetrical linear order.
2. Unilateral, continued linear order.

The headline, which is an arc, i.e., a portion of circumference, has, like the straight line, a middle or centre of figure, extremities and symmetrical subdivisions, or further, like the straight unilateral line, a starting and arriving point. Form-envelope is either an even or an uneven figure, open or close, like rings, garlands, wreaths, &c.

CHAPTER II.

PERIPHERICAL DISPOSITIONS.

154. This class contains all the dispositions brought in contact with outlines of closed figures. These outlines are polygonal, or undulated, and cur-vital dispositions, particularly polygons. (1) Quadri-laterals, rectangles, lozenges and squares have four summits, four centres, and subdivisions symmetrically placed between the middles and the summits, or between the summits only if the sub-divisions are in uneven number. The rectangle, according to pro-portion of its sides, gives rise to varied dispositions. We may, for example, consider only the four summits, the four middles, or only the two great sides, or we may adopt a sub-division into three parts for the great sides, and the middle for the small, &c. (2) Tri-gons, squares, pentagons, and generally all centred polygons, have summits, and middles, sub-divisions it may be of the sides.

The symmetrical conditions of polygons, or the relative proportions of the sides in the rectangle, limit the choice which we can make in the different varieties of repartition, the proportions of those re-partitions having to be adjusted to the symmetry which agrees with the relative proportion of the sides.

CHAPTER III.

ORBICULAR DISPOSITIONS.

155. These dispositions are referred to the con-tinued curvilinear outlines, and particularly to the ovals and circles. They have a real or imaginary head-centre which governs the outline; and they have not angular summits.

Marginal or diametrical ranges, instead of being constructed by proportion of the straight headline, can be by circular headline, i.e., a circumference. Continued marginal ranges go either way; straight marginal ranges have their axes directed externally, or from the centre of circumference. Continued diametrical, even or alternate, diagonal or contradicted ranges go either way. Straight, quartered, alternate or centred diametrical ranges have their rectilinear headline bent, following the circumference, and the axis perpendicular to this headline coinciding with rays. Starting immediately from a circumference, we can virtually find 1, 2, 4 . . . 8 . . . middles; we can also suppose 3, 6, 12 . . . or more. The case of only one middle is a ring; two middles are directly opposite, and placed at the extremities of the diameter; four middles at extremities of two crossed diameters; six middles at the extremities of three diameters or six rays, and so on. We can also divide the circumference in uneven numbers of middles, 3, 5, 7 . . . &c.; only one or two different middles for the arc, two or four different ones for the oval, are the first division adapted, naturally and simply, to the constitutive symmetry of these figures.

CHAPTER IV.

RADIATED DISPOSITIONS.

156. This class contains all the dispositions at

a head-centre, which has four form-envelopes, regular polygons and circles. It is necessary to distinguish two principal varieties :—

1. Forms being directed from the centre to the circumference, i.e., the starred dispositions.

2. Composite forms being declined in size from the circumference to the centre, i.e., dispositions and roses.

All composite forms are most naturally of even symmetry; if these forms are of uneven symmetry they will give a particular disposition, a revolved disposition, which will be the result of the over-lapping of even forms.

In spite of analogy, which brings into contact linear, circular, orbicular, irradiated and radiated ranges, we should consider these as having a superior principle of unity in a circle : (1) centre, (2) plane surface or circle, (3) circular limit or circumference, (4) rays which go from centre to circumference, (5) circular line, independent of space, which it limits on both sides. All these peculiarities, being considered singly or together, determine a number of centred forms: starred, roses, orbicular, revolved dispositions, &c.

CHAPTER V.

PENNATE DISPOSITIONS.

157. These are directed by proportion to a head-line, which can be straight, curved or turned. Composing forms of this kind are placed on both sides, articulated in successive points of the headline, and corresponding on both sides by straight, opposite and alternate symmetry. These pennate forms are articulated under angles which are, generally, acute, and which determine a principle of general unity; if the angles were straight they would have a quartering centre which breaks the unity, because it introduces two horizontal and vertical axes where the vertical line prevails.

One portion of even, straight, alternate, diametrical range can be considered as a pennate disposition. There is not, however, the declination of size of compound forms; it is an essential declination which gives unity to disposition by making it finished though it seemed indefinite.

Then, pennate disposition is determined by one headline, one form-envelope and the principle of declination, which has reference to size of forms, to their inclination, and to their form and their accentuation.

CHAPTER VI.

PALMATED DISPOSITIONS.

158. Palmated dispositions have a point or headcentre and a headline which co-exist. The form-envelope changes with co-ordination of disposition, and with co-ordination of symmetry. It is necessary to distinguish palmated dispositions from angular desinences, and full inside palmated dispositions from inside desinences and full outside dispositions. Palmated dispositions are even or uneven. In the first case the headline is an axis, the starting-point a radiating centre, and the figure-envelope a symmetrical outline. The figure-envelope is, generally, a handle or oval. It is necessary to distinguish three principal varieties in this class of dispositions : (1) Digitate disposition, when all composing forms are of the same form, gener-ally rectilinear; declination is only relative to the respective size of forms. (2) Palmated disposi-tions, properly so called, where declination is in size, then inclination of accent and variation in form of integral parts. And (3) Palmated dispositions where composing forms are continued under a radiating centre; in this case the point of insertion is inside the figure-envelope, which is either poly-gonal or oval, or rather an open arc or handle.

Two other varieties are also contained in this order :—

1. Uneven palmated dispositions, which have no rectilinear symmetrical axis, but an imaginary head-line, on both sides of which the forms follow the same or two different declinations.

2. Spread disposition or acanthus, which has a symmetrical axis and imaginary centre, if it is the diagram of every acanthus of ornamentation.

CHAPTER VII.

CURVED DISPOSITIONS.

159. Curved dispositions come from pennate and palmated dispositions; they are realized by articula-tion of forms following a curved headline. These dispositions have neither symmetry nor centre. By supposing a real curved line, this headline subordinates the forms on either side, or both at the same time; and the declination being traced from starting-point to summit there are six principal dispositions, as:—

1, 2. Forms articulated in convexity of the headline, and declined in one way or the other.

3, 4. Forms articulated in concavity of the headline, and declined in one way or the other.

5, 6. Forms articulated on both sides of the headline, and declined in one way or the other.

According as the curve is rolled more or less, they have curved, long, hooked and round dispositions.

CHAPTER VIII.

BRANCHED DISPOSITIONS.

160. Starting from declined polygonal delineations, or from a single touch, and symmetrically repeating the successive branches, we get branched dispositions, very varied, simple or compound, following the mode of symmetry and uniform number of successive parts. To this class of dispositions we can add the nerves of leaves or subdivision of composed leaves, plane figurations of trees, &c. The essential unity of this class of dispositions consists in declination of size or number of integral parts, which are placed one on the other. This declined ramification has, however, an actual end, though we may conceive an indefinite suc-

cession. In a word, it is necessary to distinguish
the physical from the theoretical termination.
Closed figures, like polygons, circles, ovals, have
theoretical and physical limit.

CHAPTER IX.

ASSEMBLED DISPOSITIONS.

161. The essential unity of these dispositions is
in the diagram, forms generally being distinct and

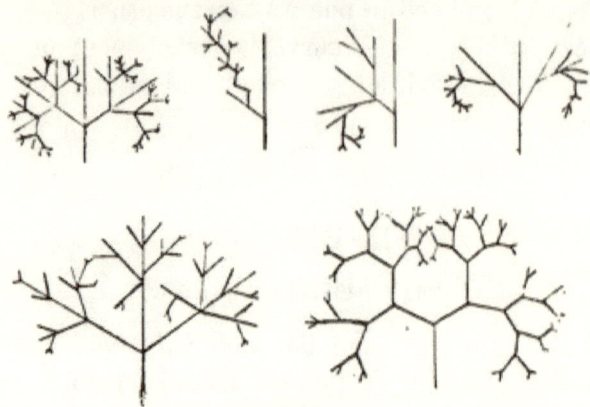

Fig. 240.

isolated one from the other. The unity of form
surpasses the unity of grouping, and generally remains distinct. That, however, does not stop the
articulation of parts, the one in proportion to the
other, or altogether in proportion to a figurative
diagram. Groupings of leaves on the branches
belong to this class.

162. Combined by syncretism, combination, conjugation, articulation,&c., these important definitions

determine a number of dispositions which belong to ornaments and natural objects. For example, take palmated dispositions. This class of dispositions contains six varieties :—

1. Turned palmated dispositions, where all developed parts, following a curved line, have their points of intersection or excision very near or mixed. They are uneven, or unsymmetrical palm leaves.

2. Radiated palmated dispositions have all their parts articulated on one axis, more or less long, but with proportion more or less strong, as in the nerves of many leaves.

4. Branched palmated divisions have their parts successively separated one from the other; the articulation takes place successively instead of being mixed in one point.

5. Assembled palmated dispositions have their parts separated, contiguous or placed over, but articulated in one point, like the groups of leaves in a branch of laurel.

6. The palmated dispositions of palm leaves.

CO-ORDINATED DISPOSITIONS.

163. This order contains all dispositions referring to the superficial extent of plane surfaces and
superfices of solid definite forms.   All plane surfaces
with rectilinear outline, without radiating centre,
such as rectangles, lozenges, squares, &c., are
limited portions of multiplied dispositions.   Poly
gons with circular dominant lines have outlines,
radiating axes, inscribed and derived figures, with a
proper centre, which governs all the system.
The diagrams of polygons are analogous to the
diagrams of circles; as :—
1. Centred network.
2. The trellis networks.
In all these networks we can put the diagonal of
meshes, which forms three sorts of rows :—
1. Circular rows.
2. Radiated rows.
3. Spiral or diagonal rows.
Such are the essential foundations of disposition
co-ordinated of circles and regular polygons.
164. Generally, all the rounded and regularly-
constructed forms have lines of curvity or repartition,
which are joined to them.   These lines of repartition
are the same as those of the plan, but changed by

continuity or curvity inherent in the corporeal form. Let us make use of these considerations with a few kinds of surfaces.

1. A cylinder is capable of development; its surface being the portion of a plan, its lines of repartition must be of three kinds—vertical, horizontal, and oblique or diagonal. It results that the dispositions in series on the cylinder are:—

    Rows following straight lines.

    Rings following circles.

    Spirals following spiral lines.

2. A cone is capable of being developed, and its surface is a portion of a circle; the lines of repartition are convergent straight lines, circles, and spiral declined lines. The divided dispositions are the same as those of circles.

3. Spheroidal surfaces. A sphere is not capable of being developed, but its curvity being uniform, the lines of repartition determine uniform networks like those in the plan. The sphere, like polyhedrons, cannot be an æsthetical form.

4. Inflected surfaces. The lines of partition are identical with the preceding; the only difference consists in those lines which are abstract imaginations, having opposite curvity.

165. All the indefinite dispositions of the first class have their analogies in every disposition that we can apply to forms or definite surfaces, as well as to plain networks; we can distinguish in these repartitions crossing-points, lines and meshes, and the ornamentation of surfaces is the same as plain ornamentation.

CONSTRUCTED DISPOSITIONS.

166. The disposition of this order results from juxtaposition of definite solid elements, whether these may have a determined form referring to the grouping in which they occur or not ; or, on the other hand, the co-ordinated disposition may result from the grouping of distinct elements.

Generally, the agglomerations, groupings and constructions of a number of objects or materials, brought in contact, determine definite disposition, according to one or several of the following conditions :—

1. The number of objects is limited.

2. The objects have definite forms, and are correlative one with the other.

3. A headline determines the grouping or organization.

4. The will and power of man, put in work, juxtaposes materials.

The arrangement of foliages, branches, bouquets, garlands, wreaths, &c., is contained in this order of dispositions.

part IV.

DEVELOPMENTS AND REFERENCES.

CHAPTER I.

APPLICATION OF QUADRIPARTITIONS.

167. We shall see that repetition determines ranges, separated, contiguous or intersected, and abatement determines repartitions, separated in line, contiguous, intersected, figured, &c., i.e., the map of ornament where the forms are co-ordinated following the harmonical lines of plan.

168. Range 1 (see Fig. 241) is obtained by the followed repetition of a quartered form.

Range 2 is obtained by the contradicted repetition of quartered form.

Range 3 is obtained by the followed repetition of even form.

Range 4 is obtained by the backward repetition of even form.

Range 5 is obtained by contradicted repetition of even form.

Range 6 the same as 5.

Range 7 is obtained by followed repetition of revolved form.

Range 8 by backward repetition of revolved form.

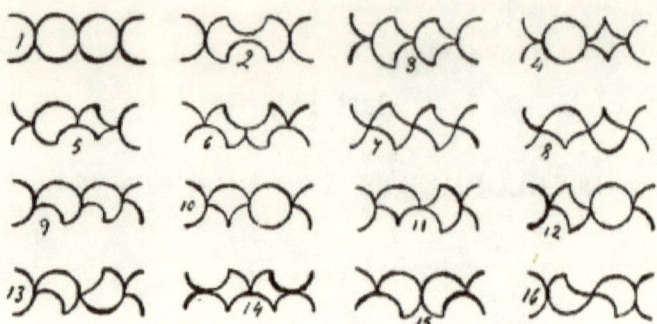

Fig. 241.

Range 9 by followed repetition of uneven form.

Range 10 by backward repetition of uneven form.

Range 11 by contradicted repetition of uneven form.

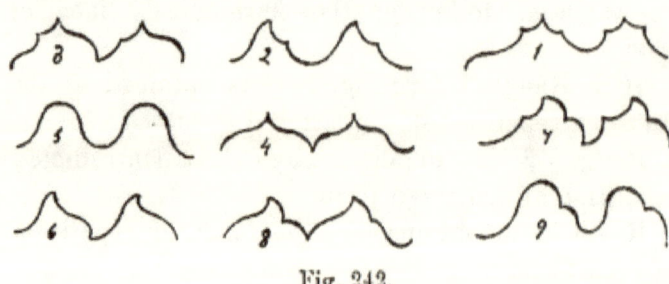

Fig. 242.

Range 12 by diagonal repetition of uneven form.

Range 13 by contradicted repetition of uneven form.

Range 14 (see Fig. 241) is obtained by alternation of two quartered and even forms.

Range 15 the same as 14.

Range 16 by alternation of quartered and revolved forms.

169. These ranges are the result of the repetition of close forms of the sixth class.

By following the continued delineations, it is easy to see that the ranges are intersecances, determined by the conjugation of the forms shown in Fig. 242.

Denticulation 1 (Fig. 242) is even marginal, wherefrom are two figures or positions.

Fig. 243.

Denticulations 2 and 3, 8 and 9, are uneven marginal, wherefrom are four positions.

Denticulations 4 and 5 are diametric alternate, wherefrom are two positions for each of them.

Denticulation 6 is diametric diagonal, wherefrom are four positions.

Denticulation 7 is diametrical contradicted, wherefrom, are four positions.

From these figures, which combine two and two, and are intersected, we may determine a number of ranges, in which are contained the preceding examples.

These denticulations are themselves determined by the conjugation and repetition of the small number of forms shown in Fig. 243.

These forms conjugated two and two by juxtaposition of squares which inscribe them, determine sixty-four conjugations, thirty-two separated and thirty-two continued. These thirty-two conjugations, contiguous, are reduced to the ten forms given in Fig. 244. These ten figures form, by the followed repetition, seven denticulations (Fig. 242). We say seven, because four of these forms are mixed two and two in one range; they are 5 and 7, and 6 and 8. These ten figures, mixed in every possible way, should give a considerable number of—(1) forms by intersecance; (2) denticulations by conjuga-

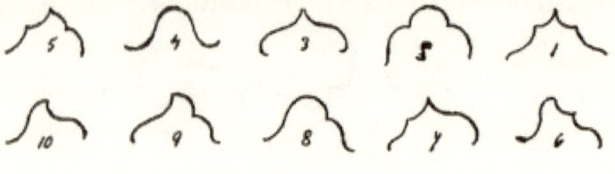

Fig. 244.

tion and repetition; (3) ranges by conjugation, repetition and intersecance; (4) repartitions by repetition, abatement, conjugation and intersecance.

REPARTITIONS.

170. Repartitions are obtained by multiplied abatement of a form, abatement which can be followed, returned, diagonal, or contradicted, separated, contiguous or intersected. The repartitions can be varied by conjugation and combination three and three, four and four, &c., of different forms. Let us take, for example, the maps in Figs. 245 to 249.

Map 1 (Fig. 245) is obtained by followed repetition of a quartered form and the straight abatement of range 1 (see p. 195, and Fig. 241).

Map 2 by followed and turned abatement of range 4.

Map 3 by followed and turned abatement of range 3.

Map 4 by followed abatement of range 5.

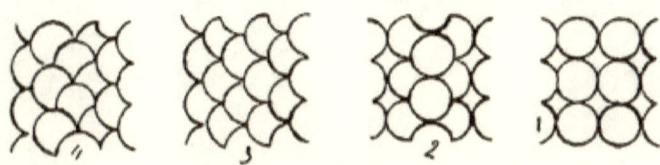

Fig. 245.

Map 5 (Fig. 246) is obtained by followed abatement of range 6 or 8 (Fig. 241).

Map 6 by returned abatement of range 6.

Map 7 by followed abatement of range 7.

Map 8 by followed and turned abatement of range 8.

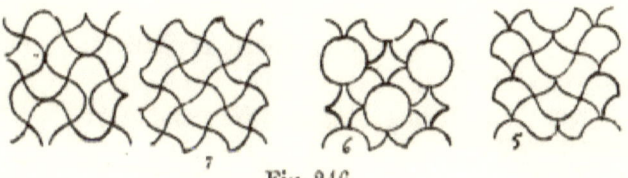

Fig. 246.

Map 9 (Fig. 247) is obtained by followed abatement of range 10 (Fig. 241).

Map 10 by followed alternate abatement of range 11.

Map 11 by followed alternate abatement of range 12.

Map 12 by returned or alternate abatement of range 13.

Map 13 (Fig. 248) is obtained by abatement of range 14 (see Fig. 241).

Map 14 by abatement of range 15.

Map 15 by abatement of range 16.

Map 16 by alternate abatement of range 16.

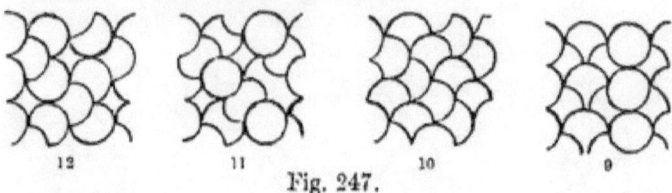

Fig. 247.

The repartitions being conjugated two and two by intersecance, we should get, for example, Fig. 249, obtained by intersecance of repartitions 6 and

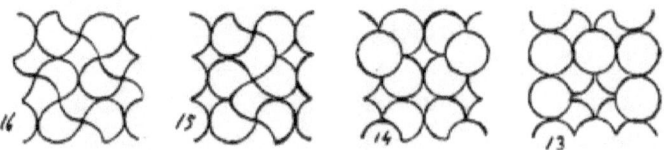

Fig. 248.

13 ; or on the contrary, these repartitions could be divided into two, as No. 7, which is composed of two repartitions (Figs. 250 and 251). The

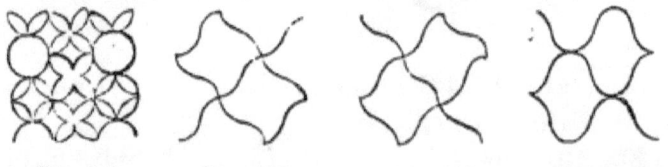

Fig. 249.   Fig. 250.   Fig. 251.   Fig. 252.

figures 250 and 252 conjugated by intersecance, produce repartition No. 16.

171. The theory of combinations has all the characteristics of a science, positive or purely rational ; it requires simple first materials, rigorously finished, and which are immovable in the different moods of construction or co-ordination to which they are exposed. According to the nature of combined elements these solutions remain distinct, or in part mixed, or exposed to a choice which, keeping the intersecting solutions, leaves the others. But this choice is not dependent on the idea of combination ; it is subordinated, on the contrary, to appreciations which refer to great notions of order and form.

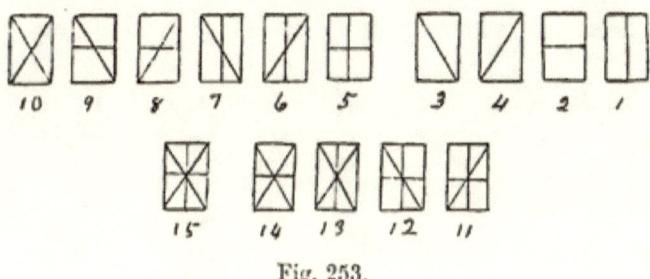

Fig. 253.

CHAPTER II.

PARTITIONS OF THE COAT OF ARMS.

172. The four partitions or principal divisions of the Coat of Arms are the only lines really simple of rectangular figures, because they are alike in their symmetrical axes. The four partitions being designed by letters as A, B, C, D, we will try to find all their possible combinations. If all these combinations were realizable, the four partitions would give rise to sixty different figures ; but really these figures are reduced to eleven only (Fig. 253),

six figures for the combinations two and two, 5,
6, 7, 8, 9, 10; four figures for the combinations
three and three, 11, 12, 13, 14; and one figure for
the combination four and four, 15. Only three of
these partitions have a proper name; No. 5 is
quartered, No. 10 is quartered crosswise, No. 15 is
lap.

CHAPTER III.

IMBRICATIONS.

173. This ornamentation, which we find every-
where, is particularly realized by the super-

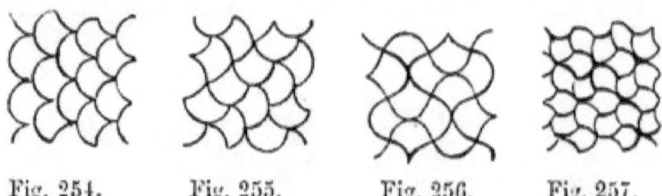

Fig. 254.   Fig. 255.   Fig. 256.   Fig. 257.

position of leaves, scales, tiles, &c., distributed
regularly and alternating. The imbrication is
uniform, being brought into contact with the
network of the plan; it is declined regularly
when brought into contact with the network
of circles, and in a varied manner when brought
into contact with the network of an ovoid.
The three rows existing conjointly, have for
the scheme the network apart from the hexagon, which has three files, and which
implicates centred composing elements instead
of even ones. The figures 254 to 256 are com-

posed of the same elements repeated for each of
them in different directions. In Fig. 254 all the
elements are directed in the same way; it is the
most general imbrication and the best known.
The arc can be a half-circumference, a raised
handle, generally even, closed or open form. In
Fig. 255 the form is repeated in two different
directions, and in Fig. 256 the form is repeated in
four directions. By supposing a mobility in the whole
system, a square network can be flattened vertically
or transversely and transformed into lozenge. By
this means of moving—(1) all forms in Fig. 254
are lengthened or flattened, and the imbrication
remains always of regular form ; (2) the forms in
Fig. 255 are lengthened by ranges ; (3) in Fig. 257
horizontal or vertical forms are consecutively
lengthened or flattened with the others, which are
also lengthened or flattened, symmetry remaining
always quartered.

174. In general, the imbrication being joined
with the condition of a complete juxtaposition, the
primary form is not arbitrary and has a regular and
definite construction. According as the lozenge
is isosceles, trigonal, or square, and the rudimental
segment even or uneven, we get forms of imbrication as follows :—1, the segment being even
or uneven, or varied in form, and the lozenge
being isosceles, trigonal, or square, we get the
form of fundamental imbrication, or that of
one direction (Fig. 254); 2, the segment being
even, simple or varied in form, and the lozenge
being isosceles, trigonal, or square, we get the
second form of imbrication, the one of two direc-

tions (Fig. 255); 3, the segment being even, simple or varied, and the lozenge being a square, we get the third form of imbrication, the one of four directions (Fig. 256); and 4, the segment being even or uneven, simple or varied, and the lozenge being trigon, we get the fourth form of imbrication, the one of three directions (Fig. 257).

This form is analogous to the third one, and can be conceived as constructed by the successive and return abatement, in two directions of a revolved form inscribed in the square. The three forms of imbrication, with one, two, and three directions, are corresponding to the three moods of juxtaposition of lozenges.

CHAPTER IV.

COMPARTMENTS AND ADJUSTMENTS.

175. All the denticulations, varied as they are, are reduced to three fundamental linear elements: uneven touches, even touches and diagonal touches, whose forms or figures can be very different. These touches succeed regularly in a straight line and follow the different moods of conjugation to determine the denticulations. The crossing of these denticulations, following the lines of networks, determines trilateral or quadrilateral meshes, whose outlines are formed by fundamental touches. Taking, for example, 1, of even form, an arc which would be the fourth part of a circumference, and inscribed, following the diagonal of a square; 2, of

uneven form, a curved line inscribed in a rectangle ;
3, of a diagonal form, two arcs joined by inflection ;
then constructing the figures following the four
quarters, determined by the axes which are cut at
right angles, we obtain the following meshes :—

I. Meshes obtained from Even Form.

176. Two rounded forms are conjugated, i.e., are
implicated. The networks which result from them
have thus two sorts of meshes; the last two
forms are also conjugated.

The third form, which is diagonal, determines a
network with mesh of one kind. The fourth form,
which is even, determines network with mesh of one
kind. If the even form were oblong, i.e., inscribed
in a rectangle, it would be an uneven form.

II. Meshes obtained from Uneven Form.

177. The two first forms are of quartered symmetry, and are both conjugated in one uniform net-

Fig. 258.

work (Fig. 259, 1) analogous to one of Fig. 245, 1.
The next two forms are even, and separately
determine two varieties of imbrication with one
direction ; but they are also conjugated, and determine conjointly a network or varied imbrication,
with two kinds of meshes (Fig. 259, 2) analogous
to the figure 256. The fifth form is diagonal, and
determines a network with mesh of one kind.

The last two forms are uneven and conjugated in
the same network.   From the chosen form it immediately results that the two networks obtained are
the figures 1, 8, 7, 16, in Figs. 245 to 248.   This effect
comes from the particular conditions of the chosen
curved line; and seeing that a form is uneven only

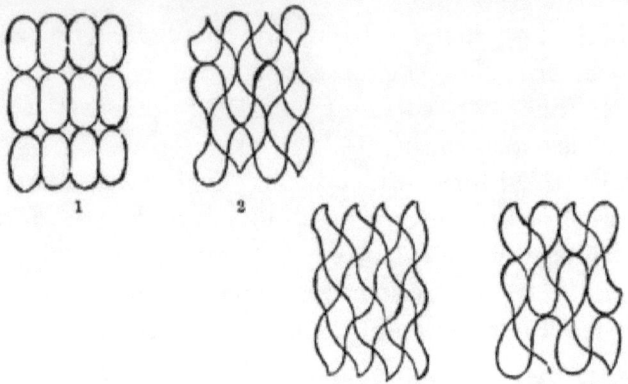

Fig. 259.

because it is neither even nor diametrical, and that
it can have the most varied forms, it is necessary to
notice here only a simple coincidence.   If the primary form were circularly disposed, we should obtain
one figure, *b* (Fig. 260), with diagonal symmetry;

Fig. 260.

two figures, *a* and *c*, with diagonal symmetry and
revolved dispositions; and three uneven figures.
The six figures which are inscribed in a square being
combined between them and with the oblong figures,
determine networks analogous to those of the
figures 245 to 249.

III. Meshes obtained from Diagonal Form.

178. The figures obtained by the quartering of diagonal touch are only four. These figures are

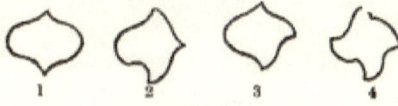

Fig. 261.

round if the touch is inscribed in a square (Fig. 261), and oblong if the touch is inscribed in a rectangle (Fig. 262).

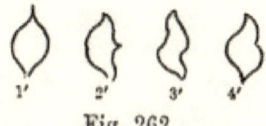

Fig. 262.

The first form 1, 1' is quartered; it determines

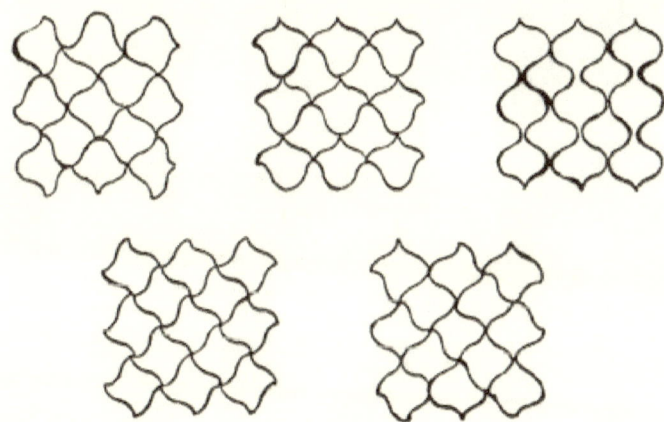

Fig. 263.

one adjustment or uniform network, composed of only one sort of mesh (Fig. 263).

The second form 2, 2' is even; it determines two

varied imbrications, one having two dimensions, the other four.

The third form 3, 3' is diagonal, and more revolved; this form determines a network composed of only one sort of mesh.

The fourth form 4, 4' is uneven; it determines one adjustment or unique network.

This network (Fig. 263) can be obtained by dividing into two the networks of Figs. 245 to 249.

All these networks are constructed on the square type. We can, by analogy, construct analogous networks on the trellis type, or following the different networks composed of varied polygons. But it would be too much to go on with such a problem. Every other diagonal form can be employed, which would give rise to very different figurations.

THE END.

Printed by GILBERT & RIVINGTON, Ld., St. John's House, Clerkenwell, London, E.C.

No. 1.—GREEK SCROLL WITH ACANTHUS LEAF.

No. 2.—Oak-leaf Bracket.

No. 3.—FLEMISH RENAISSANCE PANEL.
(*In time of Holbein.*)

No. 4.— Louis XVI. Astronomical Panel.

No. 5.—FLEMISH RENAISSANCE PANEL.

No. 6.—Panel from Nature.

No. 7.—Modern Renaissance Panel.

No. 8.—FLEMISH RENAISSANCE PANEL.

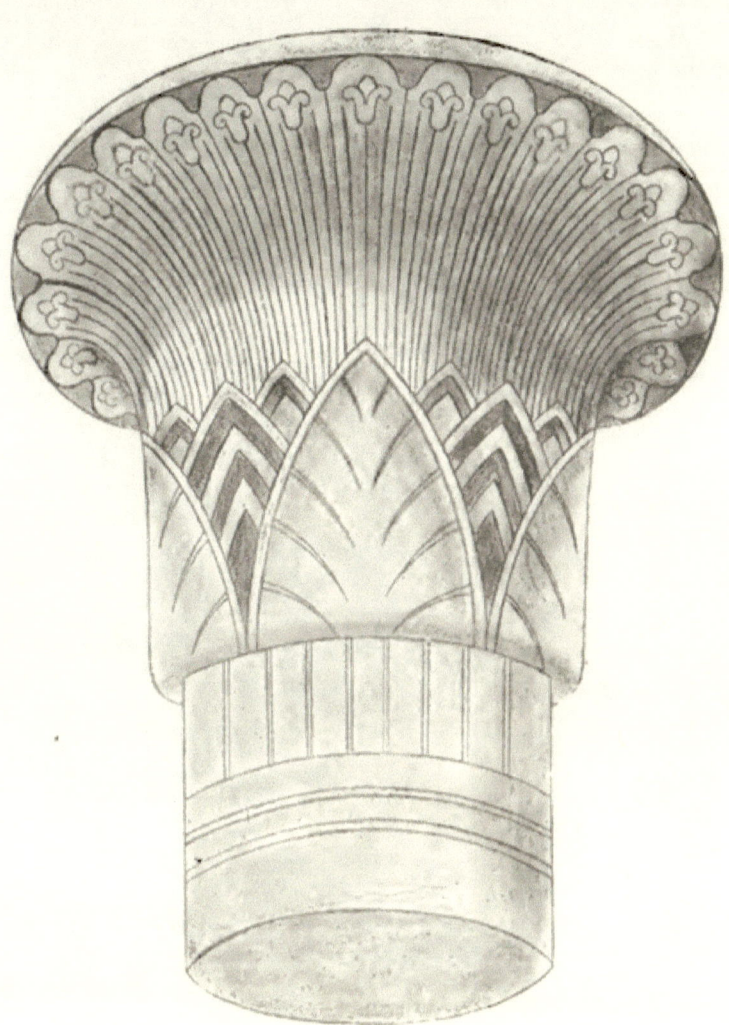

No. 9.—Egyptian Capital.

No. 10.—JAPANESE RAIL.

No. 11.—CINQUECENTO CAPITAL.

No. 12.—LOUIS XIV. PANEL.

No. 13.—JAPANESE RAIL.

No. 14.—GREEK SCROLL AND ACANTHUS-LEAF PANEL.

No. 15.—MODERN ENGLISH PANEL.

No. 16.—Italian Capital.

No. 17.—Cinquecento Bracket.

No. 18.—Greek Scroll and Acanthus Leaf (larger scale).

No. 19.—Japanese Rail (larger scale).

No. 21.—Bold Specimen of Italian Scrollwork.

No. 22.—BYZANTINE PANEL.

No. 23.—Louis XIV. Panel. (Larger scale).

No. 24.—Elizabethan Panel.

No. 25.—EGYPTIAN SCROLL PANEL.

No. 26.—Perpendicular Gothic Window.

No. 27.—MODERN ENGLISH PANEL.

No. 28.—LOUIS XVI. FLOWER PANEL.